Data Analytics

Series Editors

Longbing Cao, Advanced Analytics Institute, University of Technology, Sydney, Broadway, NSW, Australia

Philip S. Yu, University of Illinois, Chicago, IL, USA

Aims and Goals:

Building and promoting the field of data science and analytics in terms of publishing work on theoretical foundations, algorithms and models, evaluation and experiments, applications and systems, case studies, and applied analytics in specific domains or on specific issues.

Specific Topics:

This series encourages proposals on cutting-edge science, technology and best practices in the following topics (but not limited to):

- Data analytics, data science, knowledge discovery, machine learning, deep learning, big data, statistical and mathematical methods, exploratory and applied analytics,
- New scientific findings and progress ranging from data capture, creation, storage, search, computing, sharing, analysis, and visualization,
- Integration methods, best practices and typical applications across heterogeneous, multi-sources, domains and modals for data-driven real-world decision-making, and value creation.

Suggested Titles for Proposals:

- Introduction to data science
- Data science fundamentals
- Applied analytics
- Advanced analytics: concepts and applications
- Banking data analytics
- Behavior analytics
- Big data analytics
- Biomedical data analytics
- Business analytics
- Cloud analytics
- Computational intelligence methods for data science
- Data visualization
- Data optimization
- Data representation
- Distributed analytics and learning
- Educational data analytics
- Environmental data analytics
- Ethics in data science
- Feature selection and mining
- Financial data analytics and FinTech
- Government data analytics
- Health and medical data analytics
- Heterogeneous data analytics
- High performance analytics
- In-memory analytics
- Insurance data analytics
- Large-scale inference
- Learning analytics
- Large-scale learning
- Mobile analytics
- Model optimization
- Multimedia analytics
- Network analytics
- Non-IID learning
- Predictive analytics
- Prescriptive analytics
- Scientific data analytics
- Service analytics
- Smart cities, home and IoT
- Statistics for data science
- Social analytics
- Social security data analytics
- Smart city and analytics
- Spatial-temporal data analytics
- Telco data analytics
- Textual data analytics
- Time-series analysis
- Transport data analytics
- Web analytics
- Visual analytics

More information about this series at http://www.springer.com/series/15063

Youyang Qu · Mohammad Reza Nosouhi ·
Lei Cui · Shui Yu

Personalized Privacy
Protection in Big Data

 Springer

Youyang Qu (iD)
School of Information Technology
Deakin University
Melbourne, VIC, Australia

Mohammad Reza Nosouhi (iD)
School of Computer Science
University of Technology Sydney
Ultimo, NSW, Australia

Lei Cui (iD)
School of Information Technology
Deakin University
Melbourne, VIC, Australia

Shui Yu (iD)
School of Computer Science
University of Technology Sydney
Ultimo, NSW, Australia

ISSN 2520-1859   ISSN 2520-1867   (electronic)
Data Analytics
ISBN 978-981-16-3752-0   ISBN 978-981-16-3750-6   (eBook)
https://doi.org/10.1007/978-981-16-3750-6

This Springer imprint is published by the registered company Springer Nature Singapore Pte Ltd.
The registered company address is: 152 Beach Road, #21-01/04 Gateway East, Singapore 189721,
Singapore

Preface

Over the past few decades, massive volumes of data in digital form have been generated, collected, and published with the fast booming of high-performance computing devices and communicating infrastructures, which brings forward the prosperity of this big data era. Organizations, institutions, and governments are playing the key roles for collecting, storing, and sharing data. For example, social networks indicate interest and social connections of users, smart wearable devices record health status of individuals, educational institutions analyse learning patterns of students, and vehicular networks collect the daily routine of drivers. By leveraging the massive amounts of data, governments and corporations have the opportunity to improve the quality of services, bring financial benefits, and potentially create social values using diverse data processing techniques, such as machine learning, data mining, artificial intelligence, and so on. A popular real-world application scenario is that the statistics of a series of medical records is able to significantly lift the diagnosis accuracy. However, almost all collected datasets contain sensitive information implicitly or explicitly, although basic anonymization solutions have been deployed to hide the unique identifiers. Besides, the linkability of different data sources poses further challenges to privacy protection. Thus, privacy preservation has become a crucial issue that needs to be addressed in this big data age.

Personalized privacy protection is a set of emerging technologies that can personalize the privacy protection based on various indexes, such as social distance in social networks and the trade-off between privacy protection and data utility. It attracts extensive interest from both academia and industry. It can be integrated with almost all the existing mainstream privacy protection frameworks, including differential privacy, clustering-based methods, and machine learning-based models, which makes it potentially applicable in many real-world scenarios.

In this book, the target is to systematically review the state-of-the-art research of personalized privacy protection and showcase the corresponding applications. This book aims to pave the way for the forthcoming researchers, engineers, and other readers to explore this under-explored domain.

This is the first book that specifically focuses on the personalized solutions of privacy protection in big data scenarios. Most other books either mentioned personalized privacy as a future work or barely consider it as a key component. In addition to preliminary theoretical contents and conceptual explanations, this book also simplifies the interpretative procedure by jointly presenting the several corresponding applications, which are readable to both dedicated researchers and interested readers without research background in this era. The prominent and exclusive features of this book are as follows:

- Enrich understanding of the foundations and research progress of personalized privacy protection.
- Summarize the latest studies of personalized privacy protection and cover many different applications.
- Treat both advantages and disadvantages of existing personalized privacy protection techniques and share many potential research opportunities.

This monograph aspires to keep readers, including scientists and researchers, academic libraries, practitioners and professionals, lecturers and tutors, postgraduates, and undergraduates, updated with the latest algorithms, methodologies, concepts, and analytic methods for establishing future personalized privacy-preserving models and applications. It not only allows the readers to familiarize with the theoretical contents but also enables them to make the best use of the theories and develop new algorithms that could be put into practice.

The book contains roughly two main modules. In the first module, the book provides an overview of privacy protection and existing solutions, which is followed by the summary and comparison of existing leading attacks in privacy revealing area. In the second module, the book presents a series of novel models along with the corresponding application scenarios, including personalized privacy in cyber-physical systems, social networks using differential privacy, social networks using anonymity-based methods, smart homes, and location-based services. Based on the above knowledge, the book presents the identified open issues and several potentially promising future directions of personalized privacy protection, followed by a summary and outlook on the promising field. In particular, each of the chapters is self-contained for the readers' convenience. Suggestions for improvement will be gratefully received.

Melbourne, Australia   Youyang Qu
Sydney, Australia   Mohammad Reza Nosouhi
Melbourne, Australia   Lei Cui
Sydney, Australia   Shui Yu
March 2021

Acknowledgments We sincerely appreciate numerous colleagues and postgraduate students at Deakin University, Melbourne, and University of Technology Sydney, Sydney, who contribute a lot from various perspectives such that we are inspired to write this monograph. We would like to acknowledge the support from the research grant we received, namely, ARC Discovery Project under the file number 200101374. In this book, some interesting research results demonstrated are extracted from our research publications that indeed (partially) supported through the above research grants. We are also grateful to the editors of Springer, especially Dr. Nick Zhu, for his continuous professional support and guidance. Finally, we would like to express our thanks to the family of each of us for their persistent and selfless support. Without their encouragement, the book may regrettably become some fragmented research discussions.

Contents

Chapter 1
Introduction

1.1 Privacy Research Landscape

Privacy protection has been and will continue to be a long-lasting issue with the persistent data collection from various resources, such as medical institutions, social networks, government sectors, etc. The fast proliferation of smart mobile devices accelerates the data collection speed and provides sufficient storage to preserve the datasets, and thereby flourish this big data era, which poses further challenges to privacy protection.

To address the privacy issues, plenty of research has been conducted from various aspects. In classic security scenarios, there are usually three parties, including data sender, data receiver, and adversaries. Different from it, the traditional privacy protection domain assumes there are only two parties, which are data curator and data requesters, as shown in Fig. 1.1. The data curator equally treats the trustful data requesters and the adversaries. The data are transmitted to any requesters with some extent of distortion and cannot be reversed. However, the distortion should be minimized to maintain a certain level of data utility, such as statistic features.

The privacy protection methods are fast emerging in volume and dimensionality. Traditional privacy protection methods include clustering-based methods and differential privacy. The representative clustering-based methods are K-anonymity, L-diversity, and T-closeness, which focus on the volume, diversity, and distribution of each cluster, respectively. Differential privacy provides a strict definition and theoretical foundation for privacy protection. Its variants are widely deployed in various scenarios, such as the Internet of Things (IoT), social networks, machine learning, etc.

In different scenarios, new privacy-preserving techniques are experiencing a fast boom. For example, game theory based solutions are popular in the scenario where the confrontation of the data curator and the adversaries can be modeled. Federated learning, as a novel distributed learning paradigm, provides privacy-preserving model training against data island issues. Besides, to preserve the privacy of multimedia data, generative adversarial networking based solutions have proved their feasibility

Y. Qu et al., *Personalized Privacy Protection in Big Data*, Data Analytics, https://doi.org/10.1007/978-981-16-3750-6_1

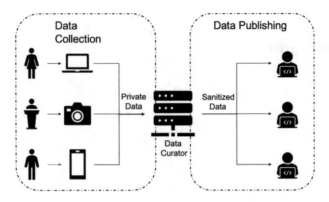

Fig. 1.1 Privacy-preserving data publishing

by several killer applications such as ZAO. There are more new techniques emerging due to the increasing demand for privacy protection in this big data era. No matter what kinds of techniques are implemented, the essential of privacy protection is to find a trade-off between privacy protection and data utility, especially for personalized privacy protection.

1.2 Personalized Privacy Overview

Traditional privacy protection usually assumes all involved parties share the same privacy protection level, which is not realistic. In real-world scenarios, the privacy protection levels should vary according to the actual demand. For example, in a social network, a piece of published data should be shown in different and personalized forms according to the intimacy of two users. Motivated by this, personalized privacy protection methods are proposed and playing an increasingly important role in the field of big data sharing.

There are two main branches for personalized privacy protection, which are static methods and dynamic methods. The static methods use one or multiple variables of the system as the index and personalize the privacy protection level based on it. These methods do not consider the interaction between the data curator (e.g., users of social networks) and the adversary. While in the dynamic methods, the data curator can adjust the privacy protection levels based on certain feedback from the adversary (e.g., attack results). This is especially important when the data is in a streaming form.

For the static methods, the general process starts from identifying a proper index, for example, the social distance in social networks. Then, a mapping function is required to map the index to a privacy protection level. At last, the design should consider the trade-off between personalized privacy protection and data utility. Optimization may be conducted in each of the processes, such as optimized indexes, constrained mapping functions, or optimal trade-offs.

Regarding the dynamic methods, the confrontation should be properly modeled. One of the most popular ways is to use game theory. For instance, by modeling the actions of the data curators and adversaries, system states, and pay-off functions, the Markov decision process can be deployed to derive the optimal privacy protection level considering the highest data utility from an overall perspective to reach the expected trade-off.

Currently, new personalized privacy protection methods are rapidly growing in volume. New theories, new techniques, and new platforms are designed to achieve a higher level of personalized privacy protection while guaranteeing optimal data utility.

1.3   Contribution of This Book

In this book, we are going to comprehensively and systematically introduce personalized privacy preserving data sharing. In this big data era, an increasingly massive volume of data is generated and transmitted, which poses great threats to privacy protection. Motivated by this, an emerging research topic, personalized privacy protection, is fast booming to serve various and diverse demands. However, there is no existing literature discussing personalized privacy protection in a systematic manner such that the knowledge in this domain is fragmentary. With this book, the authors aim to sort out the clear logic of the development of personalized privacy protection, the advantages, disadvantages, as well as the future directions of this under-explored domain.

The logic of this book follows the sequence of introduction, existing privacy protection methods, leading and emerging attacks, personalized privacy protection countermeasures, future directions. The issues of existing privacy protection methods (differential privacy, clustering, anonymity, etc.), such as low data utility and unbalanced trade-off, are identified to the necessity of personalized privacy protection. Besides, the leading and emerging attacks pose further threats to privacy protection. To mitigate the negative impact, personalized privacy protection methods are discussed in detail on both the advantages and flaws. Traditional methods, such as differential privacy, cryptography, and clustering-based methods are discussed in a comparative and intersectional way versus emerging methods like federated learning and generative adversarial nets.

To better clarify, both quantitative and qualitative results are present in figures, tables, or other suitable formats to give the readers the big picture of this topic along with unique insights of common sense and technical details. With a progressive manner, the readers will gain exclusive knowledge in personalized privacy protection and be inspired to further investigate this under-explored domain. Several prerequisites, such as differential privacy and game theory, are required to make the best use of this book, which will be explained in a simplified and concise way to help the readers to adapt in.

This book could be used as a reference for forthcoming scientists, researchers, and postgraduates to have a quick idea of the foundations and research progress of personalized privacy protection as well as being a potential textbook for lecturers, tutors, and undergraduates.

1.4 Book Overview

The remainder and book overview is as follows.

In Chap. 2, we present existing privacy protection research, including clustering- and anonymity-based methods, differential privacy and its variants, cryptography methods, and machine learning driven methods.

In Chap. 3, we summarize several mainstream privacy attacks and show their workflow. The attacks are background knowledge attacks, collusion attacks, linkage attacks, structural attacks, forgery attacks, eavesdropping attacks, and Sybil attacks.

In Chap. 4, we show our representative personalized privacy research in various scenarios, which are personalized privacy in cyber-physical systems, personalized privacy in social networks using differential privacy, personalized privacy in social networks using anonymity-based methods, personalized privacy in smart homes, and personalized privacy in location-based services.

In Chap. 5, we illustrate several potentially promising future directions for the readers. The future directions are personalized privacy-preserving attribute-based encryption, personalized privacy-preserving federated learning using generative adversarial network, personalized privacy-preserving blockchain-enabled federated learning, collusion attack resistance in personalized privacy protection, and trade-off optimization between personalized privacy protection and data utility.

In Chap. 6, we summarize and conclude this book.

Chapter 2
Existing Privacy Protection Solutions

In this chapter, we outline the major developments of modern privacy study based on the survey work we have conducted [1–4]. Mainstream privacy protection techniques including anonymity, clustering-based, differential privacy, cryptography, and machine learning methods will be presented in the following sections.

2.1 Preliminary of Privacy Study

In this section, we present an overview of privacy systems, including different participation roles, anonymization operations, and data status. We also introduce the terms and definitions of the system.

In terms of participants, we can see four different roles in the privacy protection domain.

- Data generator: Individuals or organizations who generate the original raw data (e.g., medical records of patients, bank transactions of customers), and offer the data to others in a way either actively (e.g. posting photos to social networks to the public) or passively (leaving records of credit card transactions in commercial systems).
- Data curator: The people or organizations who collect, store, hold, and release the data. Of course, the released data sets are usually anonymized before publishing.
- Data user: The people who access the released data sets for various purposes.
- Data attacker: The people who try to gain more information from the released data sets with a benign or malicious purpose. We can see that a data attacker is a special kind of data user.

There are three major data operations in privacy-preserving systems.

- Collecting: Data curators collect data from different data sources.

© The Author(s), under exclusive license to Springer Nature Singapore Pte Ltd. 2021
Y. Qu et al., *Personalized Privacy Protection in Big Data*, Data Analytics,
https://doi.org/10.1007/978-981-16-3750-6_2

Table 2.1 A table of patients in a medical database

Name	Job	Gender	Age	Disease	Other
Linda	Singer	F	30	FLU	NA
Allen	Researcher	M	25	Fever	NA
...	...	...	...	...	...

- Anonymizing: Data curators anonymize the collected data sets in order to release it to public.
- Communicating: Data users performan information retrieval on the released data sets.

Furthermore, a data set of the system possesses one of the following three different statuses.

- Raw: The original format of data.
- Collected: The data has been received and processed (such as de-noising, transforming), and stored in the storage space of the data curators.
- Anonymized. The data has been processed by an anonymization operation.

We can see that an attacker could achieve his goals by attacking any of the roles and operations. In general, we can divide a given record into four categories according to its attributes.

- Explicit identifier: A unique attribute that can clearly identify an individual, such as passport ID and drive licence numbers.
- Quasi-identifier: Attributes that be used to identify individuals with a high probability by combining other information, such as gender, birthday, age, etc. In fact, different attackers will have different quasi-identifiers according to their background knowledge.
- Sensitive attributes: The expected information interested by an adversary. In general, it is difficult to predict in advance.
- Non-sensitive attributes: The information not in the previous three categories.

We provide an example as shown in Table 2.1. In the example, *name* is an explicit identifier, while *work*, *gender*, and *age* constitute a set of quasi-identifiers, *disease* is sensitive information.

2.2   Anonymity Based and Clustering Based Methods

The data clustering direction developed from the initial k-anonymity method, then the l-diversity method, and then the t-closeness. We use Table 2.1 as an example to quickly demonstrate the journey of the data clustering methods for privacy protection.

Table 2.2 An illustration of k-anonymity (k = 2)

Job	Gender	Age	Disease	Other
Artist	M	20–30	FLU	NA
Artist	M	20–30	HIV	NA
Professional	F	30–40	FLU	NA
Professional	F	30–40	Cancer	NA

In 2000, the strategy of k-anonymity protection is shown in Table 2.2, with the strategy that each record in the table is at least as identical to the other $k - 1$ records on the quasi-identifier. Thereby reducing the probability of being identifiable. As shown in the example, dancers, singers, etc. are merged into artists, lawyers, and engineers are combined into the professional occupation, and the accurate age is expressed as a range. The k value in this example is 2. In this way, the maximum probability that a patient can be identified is $\frac{1}{k}$. If the k value is large enough, patient privacy can be effectively protected. It can be mathematically described as follows.

Let $T = t_1, t_2, ..., t_n$ be a table of a data set D, $A = A_1, A_2, ..., A_m$ be all the attributes of T, and $t_i[A_j]$ be the value of attribute A_j of tuple t_i. If $C = C_1, C_2, ..., C_k \subseteq A$, then we denote $T[C] = t[C_1], t[C_2], ..., t[C_k]$ as the projection of t onto the attributes in C.

The quasi-identifier is defined as a set of non-sensitive attributes of a table if these attributes can be linked with external data sets to uniquely identify at least one individual in the data set D. We use QI to represent the set of all quasi-identifiers.

A table T satisfies k-anonymity if for every tuple $t \in T$ there exist at least $k - 1$ other tuples $t_{i_1}, t_{i_2}, ..., t_{i_{k-1}} \in T$, such that $t[C] = t_{i_1}[C] = t_{i_2}[C], ..., t_{i_{k-1}}[C]$, for all $C \in QI$.

On the other hand, we can also note that a larger k value will result in more data loss. At the same time, under the homogenous attack, the k-anonymity model cannot effectively protect the privacy of users due to the homogeneity of sensitive attributes. For example, the attacker knew that Linda was in Table 2.2 and she had cancer. Based on this background knowledge, the attacker knows that Linda is the fourth record in the table.

To overcome the shortcomings of the k-anonymity model, Machanavajjhala et al. [48] proposed the l-diversity model in 2006, requiring at least one sensitive attribute value is different in each anonymous group. In this way, the probability of an attacker can infer a certain record of private information is up to $\frac{1}{l}$. Table 2.3 provides a concrete example, where $k = 2, l = 2$.

As aforementioned, l-diversity [23] is an extension of the k-anonymity to "well represent" the sensitive attributes. In particular, there are four different interpretations of the term "well represented" as follows.

(1) Distinct l-diversity. Similar to k-anonymity, each sensitive attribute has to possess at least l distinct values in each qid group.

Table 2.3 An Illustration of l-anonymity (k = 2, l = 2)

Job	Gender	Age	Disease	Other
Artist	F	20–30	HIV	NA
Artist	F	20–30	HIV	NA
Artist	F	20–30	Cancer	NA
Artist	F	20–30	Cancer	NA

(2) Probabilistic l-diversity. The frequency of a sensitive value in a qid group is at most $\frac{1}{l}$.

(3) Entropy l-diversity. For every qid group, its entropy is at least log l.

(4) (c, l)-diversity. The frequency of sensitive values of a qid group is confined in the range defined by c (a real number) and l (in integer).

However, the l-diversity based method cannot prevent the similarity attack, as the attacker can infer the sensitive information of the user according to the sensitive familiarity value and the semantic similarity of each QI-group. In some specific scenarios, the l-diversity model may provide more background knowledge for attackers.

In order to solve the above problems, Li et al. proposed t-Closeness in 2010. The specific strategy is: for a given QI-group, ensure that the difference between its distribution and the corresponding distribution on the original data set does not exceed a certain threshold. Based on the above three models, researchers further developed some protection methods, such as (a, k)—anonymous [5], (k, e)—nonymous [6], and (e, m)—Anonymous [7], etc. However, the anonymity-based protection models require special attack assumptions, and cannot perform quantitative analysis. Therefore, it has great limitations in practical applications.

2.3   Differential Privacy Methods

Different from the data clustering strategy, the differential privacy framework [25] was proposed in 2006, which offers strong privacy protection in sense of information theory. The basic background is that an attacker may obtain expected information by multiple queries to a statistical database on top of his background knowledge of victims. The defense strategy is: for two data sets with a minimum difference, the difference between the queries on the two data sets is very limited, therefore limiting the information gain for attackers. One popular method to achieve this is adding noise to outputs.

Definition 2.1 *Differential Privacy:* A random function M satisfies ϵ-differential privacy if for every $D_1 \sim D_2$, and for all outputs $t \in P$ of this randomized function, the following statement holds:

$$P_r[M(D_1)] \leq exp(\epsilon)P_r[M(D_2)], \tag{2.1}$$

in which exp refers to the exponential function. Two data sets D_1 and D_2 are neighbours with at most one different item. ϵ is the privacy protection parameter that controls the degree of difference induced by two neighbouring data sets. A smaller ϵ leads to a stronger privacy guarantee.

We can achieve ϵ−differential privacy by adding random noise whose magnitude is adjusted according to the global sensitivity.

Definition 2.2 *Global Sensitivity:* The global sensitivity $S(f)$ of a function f is the maximum absolute difference obtained on the output over all neighbouring data sets:

$$S(f) = \max_{D_1 \sim D_2} |f(D_1 - D_2)|. \tag{2.2}$$

Two mechanisms are always utilized to satisfy the differential privacy definition: The Laplace mechanism and the Exponential mechanism. Between these two mechanisms, the Laplace mechanism achieves ϵ−differential privacy by adding noise that following Laplace distribution is more suitable for numeric outputs.

Definition 2.3 *Laplace Mechanism:* Given a function $f: D \rightarrow P$, the mechanism $M:R \rightarrow \triangle(R)^n$ adds Laplace distributed noise to the output of f:

$$M(D) = f(D) + V, where V \sim Lap\left(\frac{S(f)}{\epsilon}\right), \tag{2.3}$$

where $Lap\left(\frac{S(f)}{\epsilon}\right)$ has PDF $\frac{1}{2\sigma}exp(\frac{-\epsilon|x|}{\sigma})$, $\sigma = \frac{S(f)}{\epsilon}$ is the scale parameter. The novel algorithm developed in this paper adopts the standard Laplacian mechanism.

Lee and Clifton [28] found that differential privacy does not match the legal definition of privacy, which is required to protect individually identifiable data, rather than the how much one individual can affect an output as differential privacy provides. As a result, they proposed differential identifiability to provide strong privacy guarantees of differential privacy, while letting policy-makers set parameters based on the established privacy concept of individual identifiability. Following this research line, Li et al. [29] analyzed the pros and cons of differential privacy and differential identifiability and proposed a framework called membership privacy. The proposed framework offers a principled approach to developing new privacy notions under which better utility can be achieved than what is possible under differential privacy.

As differential privacy is a global concept for all users of a given data set, namely the privacy protection granularity is the same to all protected users, therefore it is called uniform privacy or homogenous differential privacy. In order to offer customized privacy protection for individuals, personalized differential privacy (also named as heterogenous differential privacy or non-uniform privacy) was also extensively explored [30, 42].

2.4   Cryptography Based Methods

Based on the current situations in practice, we can conclude that encryption is still the dominant methodology for privacy protection.

Cryptography can certainly be used in numerous fashions for privacy protection in the big data age. For example, a patient can use the public key of her doctor to encrypt her medical documents and deposits the ciphertext into the doctor's online database for her treatment while her privacy is strictly preserved.

With the emergence of big data, clouds are built to serve many applications due to its economical nature and accessibility feature. For example, many medical data sets are outsourced to clouds, which triggers privacy concerns from patients. The medical records of a patient can only be accessed by authorized persons, such as her doctors, rather than other doctors or people. The public key encryption is obviously not convenient if the number of authorized people is sufficiently large due to the key management issue. In this case, Attribute-Based Encryption (ABE) is an appropriate tool [8, 9], which was invented in 2004 by Sahai and Waters [10]. In the ABE scheme, a set of descriptive attributes of the related parties, such as hospital ID and doctor ID are used to generate a secret key to encrypt messages. The decryption of a ciphertext is possible only if the set of attributes of the user key matches the attributes of the ciphertext. The ABE scheme creatively integrates encryption and access control, and therefore no key exchange problem among the members of the authorized group.

The dilemma of encryption-based privacy protection in big data is: on one hand, we need to offer sufficient privacy protection for users, at the same time, we have to make the encrypted data informative and meaningful for big data analysis and public usage. As a result, we face a number of challenges as follows. One challenge is information retrieval on encrypted data. This research branch is also called searchable encryption, which boomed around the year 2000 [11, 12]. The basic idea is as follows. An user indexes and encrypts her document collection, and sends the secure index together with the encrypted data to a server that may be malicious. To search for a given keyword, the user generates and submits a trapdoor for the keyword, which the server uses to run the search operation and recover pointers to the appropriate encrypted documents.

Another challenge here is operations on encrypted objects. This research branch is named as homomorphic encryption started in 1978 [48]. In this kind of encryptions, we expect to carry out computations on ciphertext, and obtain an encrypted output. If we decrypt the output it should match the result of operations performed on the original plaintext. Mathematically, we can describe it as follows: given a message m, a key k, and an encryption algorithm E, we can obtain a ciphertext $E_k(m)$. Let f be a function, and its corresponding function is f', D_k be a decryption algorithm under key k, then an encryption scheme is homomorphic if $f(m) = D_k(f'(E_k(m)))$.

In 2009, Gentry kicked off a further development in this direction, Fully Homomorphic Encryption (FHE), which supports arbitrary computation on ciphertexts [13]. A survey on this branch can be found in [14]. The problem is that we do not have a feasible fully homomorphic encryption system in place yet due to the extraordinary inefficiency in computing. Compared to FHE, Multi-Party Computation (MPC), which was initiated by Yao in 1982 [15], has been used in practice by offering weaker security guarantees but much more efficient. The scenario of MPC is like this: multiple participants jointly compute a public function based on their private inputs while reserving their input privacy against the other participants, respectively.

We have to note that encryption can protect the privacy of an object itself, however, it is vulnerable against side information attacks, such as traffic analysis attacks against anonymous communication systems. For example, we can encrypt web pages of a protected website, however, the encryption cannot change the fingerprints of the web pages, which are represented by the size of the HTML text, number of webobjects, and the size of the web objects. An attacker can figure out which web pages or web sites a victim visited using traffic analysis methodology [16–18]. In terms of solutions, information theory based packet padding is the main player, including dummy packet padding [19] and predicted packet padding [20].

2.5 Machine Learning and AI Methods

The flourishing of machine learning (ML) has become one of the drivers of privacy concerns in modern society. Sensitive information of users may be compromised during the data collecting and model training process. Fortunately, recent studies have shown that some ML methods can also act as tools for privacy protection if employed correctly. Novel decentralized learning framework which can facilitate distributed learning tasks and enable source data to remain on edge devices has received widespread attention [21].

Distributed training system contains the following main modules: data and model partitioning module, stand-alone optimization module, a communication module, as well as data and model aggregation module. In particular, different machines are responsible for different parts of the model and assigned with different data. Therefore, distributed training systems can keep datasets containing privacy at different locations instead of the cloud, which have been widely applied in recent years.

Edge computing is a widely applied decentralized architecture that performs processing tasks in intelligent edge nodes. Similar to distributed training, the architecture of edge computing can mitigate privacy issues [22]. However, other security techniques are required to combine with. For example, Gai et al. [23] combined blockchain and edge computing techniques to address the security and privacy issues in smart grid. Ma et al. [24] proposed a lightweight privacy-preserving classification framework for face recognition by employing additive secret sharing and edge computing. Li et al. [25] proposed a privacy protection data aggregation scheme for mobile edge computing assisted IoT applications based on the Boneh-Goh-Nissim cryptosystem. Du et al. [26] handled with privacy problems of training datasets and proposed a differential privacy based protection method in wireless big data with edge computing.

Federated learning (FL), as a novel distributed learning paradigm, becomes prominent recently to address privacy issues. Different from traditional methods which put all data $D_1 \cup D_2 \cup ... \cup D_N$ to train a model M_{SUM}, a FL system is a learning process in which the data owners collaboratively train a model M_{FED}, while any data owner F_i dose not share her data D_i to others [27]. In addition, let V_{FED} represent the accuracy of M_{FED}, it should be very close to V_{SUM}. Specifically, let δ denote a non-negative real number, if

$$|M_{FED} - M_{SUM}| < \delta, \tag{2.4}$$

we say that the federated learning algorithm has $\delta-$accuracy loss.

However, recent studies have demonstrated that the framework of FL also has some privacy issues [28]. One major concern is that adversaries could recover sensitive data by violating the shared parameters. To mitigate this problem, Bonawitz et al. [29] designed a secure aggregation method to protect the privacy of each user's model gradient. Recently, Liu et al. [30] pointed out that user dropout and untrusted server are two unresolved challenges of original FL schemes. Thus, they proposed a robust federated extreme gradient boosting framework for mobile crowdsensing that supports forced aggregation. Hao et al. [31] proposed a privacy-enhanced FL scheme by employing homomorphic ciphertext and differential privacy. The proposed noninteractive method can achieve both effective protection and efficiency. From the perspective of verifying whether the cloud server is operating correctly, Xu et al. [28] proposed a privacy-preserving and verifiable FL framework based on double-masking protocol.

References

1. J. Yu, K. Wang, D. Zeng, C. Zhu, S. Guo, Privacy-preserving data aggregation computing in cyber-physical social systems. ACM Trans. Cyber-Phys. Syst. **3**(1), 1–23 (2018)
2. L. Cui, G. Xie, Y. Qu, L. Gao, Y. Yang, Security and privacy in smart cities: challenges and opportunities. IEEE Access **6**, 134–146 (2018)
3. Y. Qu, M.R. Nosouhi, L. Cui, S. Yu, Privacy preservation in smart cities, in *Smart Cities Cybersecurity and Privacy* (Elsevier, Amsterdam, 2019), pp. 75–88
4. Y. Qu, S. Yu, W. Zhou, S. Peng, G. Wang, K. Xiao, Privacy of things: emerging challenges and opportunities in wireless internet of things. IEEE Wirel. Commun. **25**(6), 91–97 (2018)
5. R.C.-W. Wong, J. Li, A. W.-C. Fu, K. Wang, (α, k)-anonymity: an enhanced k-anonymity model for privacy preserving data publishing, in *Proceedings of the 12th ACM SIGKDD International Conference on Knowledge Discovery and Data Mining* (ACM, 2006), pp. 754–759
6. Q. Zhang, N. Koudas, D. Srivastava, T. Yu, Aggregate query answering on anonymized tables, in *IEEE 23rd International Conference on Data Engineering* (IEEE, 2007), pp. 116–125
7. J. Li, Y. Tao, X. Xiao, Preservation of proximity privacy in publishing numerical sensitive data, in *Proceedings of the 2008 ACM SIGMOD International Conference on Management of Data* (ACM, 2008), pp. 473–486
8. V. Goyal, O. Pandey, A. Sahai, B. Waters, Attribute-based encryption for fine-grained access control of encrypted data, in *Proceedings of the 13th ACM Conference on Computer and Communications Security* (2006), pp. 89–98

9. A. Lewko, B. Waters, Decentralizing attribute-based encryption, in *Annual International Conference on the Theory and Applications of Cryptographic Techniques* (Springer, 2011), pp. 568–588

10. A. Sahai, B. Waters, Fuzzy identity-based encryption, in *Annual International Conference on the Theory and Applications of Cryptographic Techniques* (Springer, 2005), pp. 457–473

11. D.X. Song, D. Wagner, A. Perrig, Practical techniques for searches on encrypted data, in *Proceeding of the IEEE Symposium on Security and Privacy. S&P 2000* (IEEE, 2000), pp. 44–55

12. R. Curtmola, J. Garay, S. Kamara, R. Ostrovsky, Searchable symmetric encryption: improved definitions and efficient constructions. J. Comput. Secur. **19**(5), 895–934 (2011)

13. C. Gentry, Fully homomorphic encryption using ideal lattices, in *Proceedings of the Forty-first Annual ACM Symposium on Theory of Computing* (2009), pp. 169–178

14. V. Vaikuntanathan, Computing blindfolded: new developments in fully homomorphic encryption, in *IEEE 52nd Annual Symposium on Foundations of Computer Science* (IEEE, 2011), pp. 5–16

15. A.C. Yao, Protocols for secure computations, in *23rd Annual Symposium on Foundations of Computer Science (sfcs)* (IEEE, 1982), pp. 160–164

16. Q. Sun, D.R. Simon, Y.-M. Wang, W. Russell, V.N. Padmanabhan, L. Qiu, Statistical identification of encrypted web browsing traffic, in *Proceedings of the IEEE Symposium on Security and Privacy* (IEEE, 2002), pp. 19–30

17. M. Liberatore and B. N. Levine, Inferring the source of encrypted http connections, in *Proceedings of the 13th ACM Conference on Computer and Communications Security* (2006), pp. 255–263

18. Y. Zhu, X. Fu, B. Graham, R. Bettati, W. Zhao, Correlation-based traffic analysis attacks on anonymity networks. IEEE Trans. Parallel Distrib. Syst. **21**(7), 954–967 (2009)

19. P. Venkitasubramaniam, T. He, L. Tong, Anonymous networking amidst eavesdroppers. IEEE Trans. Inf. Theory **54**(6), 2770–2784 (2008)

20. S. Yu, G. Zhao, W. Dou, S. James, Predicted packet padding for anonymous web browsing against traffic analysis attacks. IEEE Trans. Inf. Forensics Secur. **7**(4), 1381–1393 (2012)

21. Y. Sun, J. Liu, J. Wang, Y. Cao, N. Kato, When machine learning meets privacy in 6g: a survey. IEEE Commun. Surv. Tutorials (2020)

22. J. Zhang, B. Chen, Y. Zhao, X. Cheng, F. Hu, Data security and privacy-preserving in edge computing paradigm: survey and open issues. IEEE Access **6**, 209–237 (2018)

23. K. Gai, Y. Wu, L. Zhu, L. Xu, Y. Zhang, Permissioned blockchain and edge computing empowered privacy-preserving smart grid networks. IEEE Internet Things J. **6**(5), 7992–8004 (2019)

24. Z. Ma, Y. Liu, X. Liu, J. Ma, K. Ren, Lightweight privacy-preserving ensemble classification for face recognition. IEEE Internet Things J. **6**(3), 5778–5790 (2019)

25. X. Li, S. Liu, F. Wu, S. Kumari, J.J. Rodrigues, Privacy preserving data aggregation scheme for mobile edge computing assisted iot applications. IEEE Internet Things J. **6**(3), 4755–4763 (2018)

26. M. Du, K. Wang, Z. Xia, Y. Zhang, Differential privacy preserving of training model in wireless big data with edge computing. IEEE Trans. Big Data (2018)

27. Q. Yang, Y. Liu, T. Chen, Y. Tong, Federated machine learning: concept and applications. ACM Trans. Intell. Syst. Technol. (TIST) **10**(2), 1–19 (2019)

28. G. Xu, H. Li, S. Liu, K. Yang, X. Lin, Verifynet: secure and verifiable federated learning. IEEE Trans. Inf. Forensics Secur. **15**, 911–926 (2019)

29. K. Bonawitz, V. Ivanov, B. Kreuter, A. Marcedone, H.B. McMahan, S. Patel, D. Ramage, A. Segal, K. Seth, Practical secure aggregation for federated learning on user-held data (2016), arXiv preprint arXiv:1611.04482

30. Y. Liu, Z. Ma, X. Liu, S. Ma, S. Nepal, R. Deng, Boosting privately: privacy-preserving federated extreme boosting for mobile crowdsensing (2019), arXiv preprint arXiv:1907.10218

31. M. Hao, H. Li, X. Luo, G. Xu, H. Yang, S. Liu, Efficient and privacy-enhanced federated learning for industrial artificial intelligence. IEEE Trans. Industr. Inf. **16**(10), 6532–6542 (2019)

Chapter 3
Leading Attacks in Privacy Protection Domain

In this chapter, we discuss seven leading attacks in privacy domain built upon the major privacy concerns in general cases. There are background knowledge attack, collusion attack, linkage attack, structural attack, forgery attack, eavesdropping attack, and Sybil attack. There are also some other forms of attacks such as tracking attacks [1] and inference attacks [2, 3], but these attacks fall in the range of the seven illustrated attacks. Beyond the traditional privacy protection scenarios, these attacks still reveal the privacy and may result in further privacy leakage when personalized privacy protection solutions are deployed, which will be detailedly discussed in the following chapters.

3.1 Major Privacy Concerns

With the widespread of mobile devices, massive data is being generated at every moment. The privacy protection under big data scenario has new features and development [4]. Usually, the released data contains sensitive identity information, location information, other profile information, etc [5]. Although a single piece of data usually does not cause privacy leakage, multiple pieces of data can be regarded as a combination of quasi-identifiers and may lead to intractable privacy loss. In Table 3.1, we further illustrate the correlation between privacy issues and attacks. We summarize all the privacy issues and attacks in mobile social networks as follows.

3.1.1 Identity Privacy

Protecting identity privacy [6] is the most fundamental target in privacy protection in social networks. If identity privacy is breached, most of the following sensitive

© The Author(s), under exclusive license to Springer Nature Singapore Pte Ltd. 2021
Y. Qu et al., *Personalized Privacy Protection in Big Data*, Data Analytics,
https://doi.org/10.1007/978-981-16-3750-6_3

information will leak accordingly. This can be achieved in several ways, for example, anonymity, pseudonym generation [7], and so on. The final target is to prevent adversaries from re-identifying specific users, which is essential especially in social network data sharing. In [8], Wang et al. investigated crowd-sourced data publishing in social networks using differential privacy. The investigated data is real-time as well as spatiotemporal. This work takes the continuous publication of statistics and demonstrates the "RescueDP", which is an online aggregate supervisory control framework with w-event privacy preservation. The core elements include adaptive sampling, dynamic clustering, adaptive budget allocation, filters, and perturbation. In addition, the authors developed a reinforced RescueDP based on neural networks to calculate the statistics and thereby improve data utility. Xing et al. [9] proposed a k-means-based community establishment scheme in social networks with privacy protection. This scheme maintains the privacy of both sensitive information of individual and statistics features of the community. In each iteration of k-means algorithm, the scheme processes two privacy-preserving operations. The first one is that users try to find nearest clusters without knowing the cluster centers. The second one is that the cluster centers are calculated without information leakage and users inside a specific cluster cannot infer the identity of each other.

3.1.2   Anonymization Versus De-Anonymization

Anonymization is another big issue which is closely related to identity privacy. Usually, anonymization is used for publishing the big social data sets for research or commercial purposeS [10]. The most economical way of data release is anonymization. Modern anonymization methods are far beyond simply eliminating the identifiers, for example, adding nodes or modifying edges to introduce random noise [11]. However, fast development of de-anonymization techniques [12] puts anonymization under great threats.

3.1.3   Location Privacy

Beyond identity privacy, location privacy [13] has attracted plentiful attention from researchers. As social network users spend more and more time and energy on mobile devices, mobile social applications may cause location privacy leakage by accessing the users' GPS data [14]. Adversaries can easily obtained either from the released data or from crawling it from the system background [15]. For example, a specific user may publish the location information when enjoying a fancy dinner at a restaurant or adversaries can hack the application directly. Therefore, improper release of location sensitive data can even cause physical loss.

3.1.4   Content Oriented Privacy (CO Privacy)

Mobile social networks can be regarded as a specific type of content-oriented networks while privacy protection in content-oriented networks has always been considered [16]. As discussed in [17], content-oriented privacy consists of three properties, which are immutability, transparency, and accountability.

3.1.5   Interest Privacy

In social networks, users are usually categorized by interest communities. Adversaries can launch collusion attack, background knowledge attack, or inference attack to gain interest privacy information by breaching the privacy of anyone in the community [18]. Moreover, built upon the location privacy, adversaries can obtain sensitive information, for example, favourite restaurant, preferred cinema, best-loved bookshop, and so on [19]. Based on the interest-based sensitive information, adversaries can spam users or commit other malicious attacks with potential profitable targets.

3.1.6   Backward Privacy and Forward Privacy (B&G Privacy)

Backward privacy denotes that an adversary cannot track the previous actions of users when the adversary has the sensitive information stored in it, while forward privacy is that an adversary cannot predict the previous actions of users when the adversary has the sensitive information stored in it [20]. These two features are quite important as privacy protection in social networks should always be long-term protection [15]. Thus, the sensitive information should be context-aware and the privacy protection must take the forward and backward status into consideration.

3.2   Leading Privacy Breaching Attacks

3.2.1   Background Knowledge Attack

Background knowledge attack is one of the most popular attacks under privacy scenarios. The rationale behind its proliferation is that it can be combined with other types of attacks. Background knowledge of a specific entry is easy to obtain in mobile social networks [21]. Moreover, the background knowledge of adversaries is hard to model, measure, and predict, which makes it more difficult to be defeated.

Table 3.1 Privacy issues and corresponding attacks

	Identity privacy	Anonymization	Location privacy	CO privacy	Interest privacy	B&F privacy
Background knowledge attack	✓	✓	✓	✓	✓	✓
Collusion attack	✓	✓	✓	✓	✓	O
Linkage attack	✓	✓	✓	✓	✓	✓
Structural attack	✓	✓	O	×	×	O
Forgery attack	O	O	✓	×	×	O
Eavesdropping attack	✓	✓	✓	✓	✓	O
Sybil attack	✓	✓	O	×	×	O

✓ denotes fully supported; × denotes not supported; O denotes partially supported

3.2.2 Collusion Attack

Collusion attack is another wide-spread attack method. Collusion attack is especially mortal in mobile social network circumstances. The reason is that a specific user can have multiple contacts in social networks and therefore there might be multiple adversaries hiding in the contact list. As different adversary holds different background knowledge of this user, they can share the information with each other to launch a collusion attack [22]. In addition, collusion attack can also be combined with other forms of attacks.

3.2.3 Linkage Attack

Linkage attack is experiencing rapid expansion with rapidly increasing data volume and data sources. For example, adversaries can make an attack based on multiple social networks. Linkage attack has a good attack performance as adversaries can collect different category of data of the same user from multiple data sources [23]. Furthermore, machine learning-based methods provide linkage attack better tools which help adversaries bypass the protection. Song et al. [24] developed a new type of inference attack. This type of attack targets on the browsing history of Twitter users leveraging twitter metadata and public click analytics. This attack only needs Twitter profile information and URL shortening services, which are public and easy-to-access information. This can further reduce the attack overhead and upgrade accuracy by taking time-varying models of users into consideration.

3.2.4 Structural Attack

Adversaries are proceeding to structural attacks because social networks are usually modelled as a graph based on graph theory. In one hand, graph theory helps to better understand and establishes social networks structure. On the other hand, adversaries can take advantage of the structural information to mount an attack. The most outstanding merit of structural attack is that adversaries can re-identify a specific user even without background knowledge. The structural attack is also widely-deployed in de-anonymization. In [25], Chen et al. proposed two types of practical attacks to steal sensitive information from graph-based clustering methods. Targeted noise injection and small community are devised to attack three popular graph clustering models, including community discovery, node2vec, and singular value decomposition (SVD). Based on this, the authors found that adversaries with limited open-source background knowledge can launch successful attacks. In term of simple defenses, it can decrease the success ratio to 25% by the cost of only 0.2% clusters over-noisy.

3.2.5 Forgery Attack

In a forgery attack, misleading messages are generated with fake information, so that adversaries can initiate some other plotting attacks such as the location-tracking attack. There are five phases in a forgery attack, in which we use vehicular social networks as an example. Firstly, the victim node and the adversary node establish a link with location information. Secondly, the adversary node creates malicious payload to the victim node. Thirdly, the victim node sends a request to a social spot s_1 for cookies. Fourthly, s_1 gives the email address of victim node to the adversary node. Lastly, the social spots reset the certificates. In this way, an outside forgery attack is performed and the privacy of the victim nodes will be compromised [26].

3.2.6 Eavesdropping Attack

In the case of eavesdropping attack, it is quite intuitive that adversaries eavesdrop the information communication and transmission process by means of modern hacking technologies, including internet, electromagnetic wave, and so on. This type of attack is launched by unauthorized real-time interception of a private communication [27]. Therefore, it is vital to secure communication to prevent privacy leakage.

3.2.7 Sybil Attack

The Sybil attack is normally launched under the scenario of a reputation-involved system. During the attack process, an adversary generates a large number of pseudo names and further gains the maximum influence [28]. Based on the influence, the adversary can mislead the other users in the system or even fool the central authority. Privacy leakage happens during the attack. Whether the attack can be successfully launched is decided by the cost to fake identities and the trust mechanism between central authority and the identities. In [29], Liu et al. did a study on extended Sybil defences. The authors found that current sybil attack models in social networks are static, which is not practical. This work takes temporal dynamics into consideration and involves three new features. Firstly, the new model considers the capabilities of adversaries to modify Sybil-controlled parts of a structural social graph. Secondly, another new feature is the capabilities to modify the connections which Sybil identities of him/her maintain to honest users. Thirdly, the proposed model benefits from the regular dynamics of connections structure and thereby trains social networks' honest parts.

References

1. H. Xu, S. Hao, A. Sari, H. Wang, Privacy risk assessment on email tracking (2018)
2. M. Nasr, R. Shokri, A. Houmansadr, Comprehensive privacy analysis of deep learning: passive and active white-box inference attacks against centralized and federated learning, in *IEEE Symposium on Security and Privacy (SP)* (IEEE, 2019), pp. 739–753
3. B. Mei, Y. Xiao, R. Li, H. Li, X. Cheng, Y. Sun, Image and attribute based convolutional neural network inference attacks in social networks. IEEE Trans. Netw. Sci. Eng. (2018)
4. S. Yu, Big privacy: challenges and opportunities of privacy study in the age of big data. IEEE Access **4**, 2751–2763 (2016)
5. C. Wu, X. Chen, W. Zhu, Y. Zhang, Socially-driven learning-based prefetching in mobile online social networks. IEEE/ACM Trans. Netw. **25**(4), 2320–2333 (2017)
6. A. Martínez-Ballesté, P. A. Pérez-Martínez, A. Solanas, The pursuit of citizens' privacy: a privacy-aware smart city is possible. IEEE Commun. Mag. **51**(6) (2013)
7. H. Liu, X. Li, H. Li, J. Ma, X. Ma, Spatiotemporal correlation-aware dummy-based privacy protection scheme for location-based services, in *Proceedings of IEEE INFOCOM Conference on Computer Communications, INFOCOM 2017, Atlanta, GA, USA, May 1–4, 2017* (2017), pp. 1–9
8. Q. Wang, Y. Zhang, X. Lu, Z. Wang, Z. Qin, K. Ren, Real-time and spatio-temporal crowd-sourced social network data publishing with differential privacy. IEEE Trans. Dependable Sec. Comput. (2016)
9. K. Xing, C. Hu, J. Yu, X. Cheng, F. Zhang, Mutual privacy preserving k-means clustering in social participatory sensing. IEEE Trans. Industr. Inf. **13**(4), 2066–2076 (2017)
10. S. Ji, P. Mittal, R.A. Beyah, Graph data anonymization, de-anonymization attacks, and de-anonymizability quantification: a survey. IEEE Commun. Surv. Tutorials **19**(2), 1305–1326 (2017)
11. K. Liu, E. Terzi, Towards identity anonymization on graphs, in *Proceedings of the ACM SIG-MOD International Conference on Management of Data, SIGMOD 2008, Vancouver, BC, Canada, June 10–12, 2008* (2008), pp. 93–106

12. S. Ji, W. Li, M. Srivatsa, J.S. He, R.A. Beyah, General graph data de-anonymization: from mobility traces to social networks. ACM Trans. Inf. Syst. Secur. **18**(4), 12:1–12:29 (2016)
13. A.R. Beresford, F. Stajano, Location privacy in pervasive computing. IEEE Pervasive Comput. **2**(1), 46–55 (2003)
14. T. Shu, Y. Chen, J. Yang, A. Williams, Multi-lateral privacy-preserving localization in pervasive environments, in *2014 IEEE Conference on Computer Communications, INFOCOM 2014, Toronto, Canada, April 27 - May 2, 2014* (2014), pp. 2319–2327
15. W. Wang, Q. Zhang, Privacy preservation for context sensing on smartphone. IEEE/ACM Trans. Netw. **24**(6), 3235–3247 (2016)
16. A. Chaabane, E. De Cristofaro, M.A. Kaafar, E. Uzun, Privacy in content-oriented networking: threats and countermeasures. ACM SIGCOMM Comput. Commun. Rev. **43**(3), 25–33 (2013)
17. P. Zhang, Q. Li, P.P.C. Lee, Achieving content-oriented anonymity with CRISP. IEEE Trans. Dependable Sec. Comput. **14**(6), 578–590 (2017)
18. X. Wang, X. Luo, S. Zhang, Y. Ding, A privacy-preserving fuzzy interest matching protocol for friends finding in social networks, in *Springer Software Computing* (2017)
19. J. Zhou, Z. Cao, X. Dong, X. Lin, Security and privacy in cloud-assisted wireless wearable communications: challenges, solutions, and future directions. IEEE Wirel. Commun. **22**(2), 136–144 (2015)
20. D. He, S. Zeadally, An analysis of RFID authentication schemes for internet of things in healthcare environment using elliptic curve cryptography. IEEE Internet Things J. **2**(1), 72–83 (2015)
21. D. Riboni, L. Pareschi, C. Bettini, Js-reduce: defending your data from sequential background knowledge attacks. IEEE Trans. Dependable Sec. Comput. **9**(3), 387–400 (2012)
22. M. Rezvani, A. Ignjatovic, E. Bertino, S. Jha, Secure data aggregation technique for wireless sensor networks in the presence of collusion attacks. IEEE Trans. Dependable Sec. Comput. **12**(1), 98–110 (2015)
23. R. Yu, J. Kang, X. Huang, S. Xie, Y. Zhang, S. Gjessing, Mixgroup: accumulative pseudonym exchanging for location privacy enhancement in vehicular social networks. IEEE Trans. Dependable Sec. Comput. **13**(1), 93–105 (2016)
24. J. Song, S. Lee, J. Kim, Inference attack on browsing history of twitter users using public click analytics and twitter metadata. IEEE Trans. Dependable Sec. Comput. **13**(3), 340–354 (2016)
25. Y. Chen, Y. Nadji, A. Kountouras, F. Monrose, R. Perdisci, M. Antonakakis, N. Vasiloglou, Practical attacks against graph-based clustering, in *Proceedings of the 2017 ACM SIGSAC Conference on Computer and Communications Security, CCS 2017, Dallas, TX, USA, October 30 - November 03, 2017* (2017), pp. 1125–1142
26. M.A. Ferrag, L.A. Maglaras, A. Ahmim, Privacy-preserving schemes for ad hoc social networks: a survey. IEEE Commun. Sur. Tutorials **19**(4), 3015–3045 (2017)
27. B. Ying, D. Makrakis, H.T. Mouftah, Privacy preserving broadcast message authentication protocol for vanets. J. Netw. Comput. Appl. **36**(5), 1352–1364 (2013)
28. D. Quercia, S. Hailes, Sybil attacks against mobile users: friends and foes to the rescue, in *Proceedings of INFOCOM 2010. 29th IEEE International Conference on Computer Communications, Joint Conference of the IEEE Computer and Communications Societies, 15–19 March 2010, San Diego, CA, USA* (2010), pp. 336–340
29. C. Liu, P. Gao, M.K. Wright, P. Mittal, Exploiting temporal dynamics in sybil defenses, in *Proceedings of the 22nd ACM SIGSAC Conference on Computer and Communications Security, Denver, CO, USA, October 12–6, 2015* (2015), pp. 805–816

Chapter 4
Personalized Privacy Protection Solutions

In this chapter, we demonstrate personalized privacy protection techniques that applied to secure real-world applications such as cyber-physical systems, social networks, smart homes, and location-based services.

4.1 Personalized Privacy in Cyber Physical Systems

The ubiquitous existence and fast proliferation of mobile devices and internet access accelerates the popularization of cyber-physical social networks (CPSN). The CPSN is an enhanced version of the classic social networks that map cyber space to physical world by users actively publishing data including location information on the service apps [1]. According to [2], various cyber-physical social networks (CPSNs) have been installed on over 80% smart mobile devices. Everyone can browse the published text, user nickname, and location information in CPSNs, for example, local business service system like "Groupon" or "Scoopon" [3]. In a CSPN, users act like sensors themselves and the data published is regarded as the sensing data [4], which is usually accessible to the public without proper access control and subsequently raises great privacy concerns.

The privacy leakage of sensitive information arises extensive concerns because of the proliferation of various cyber-physical social networks (CPSNs) installed on smart mobile devices. One of the greatest bottleneck of data sharing over CPSNs is privacy, in particular, customizable privacy issues [5]. The data is shared with various recipients, including adversaries. However, existing privacy protection schemes offer privacy protection of the same level, which is referred as uniform privacy protection. This causes possible leakage of sensitive information or degradation of data utility. Motivated by this, customizable privacy protection models are proposed to address the problem. They have been applied with a lot of real-world scenarios, for example,

Y. Qu et al., *Personalized Privacy Protection in Big Data*, Data Analytics, https://doi.org/10.1007/978-981-16-3750-6_4

edge computing [5, 6] and mobile crowd-sensing in Industrial Internet of Things (IIoT) [7]. However, current works focus more on data publishing of a whole data set, while privacy-preserving data sharing in CPSNs is barely discussed.

Despite the flexibility and individual-specificity offered by customizable differential privacy, it is subjected to leading attacks such as background knowledge attack and collusion attack. The rationale behind this issue is that the customization of privacy protection levels enables customized information from the recipients side, which could be used to infer or collude for further sensitive information. The customized data contains unexpected correlation of injected noises constrained by differential privacy, which is not sufficiently studied. Adversaries can leverage background knowledge or collude with each other to ceaselessly launch these two attacks since the data shared in CPSNs is updated from time to time [8]. In addition, both adversaries and attacks are measured qualitatively rather than quantitatively in most existing works [9]. In this context, the adversaries and attacks cannot be modelled where the negative effects cannot be theoretically considered while neglecting this puts customizable differential privacy protection under further threats.

In addition to privacy protection, it is equally important to consider the data utility in CPSNs. Data utility determines the functionality of CPSNs, which means users cannot be satisfied by browsing over-sanitized information [4]. Simultaneously, data analysts require the statistical regularity of the released data for research purposes such as recommendation mechanisms in CPSNs [10]. All these issues jointly generate the demand on a balance between privacy protection and data utility, which is also known as a trade-off.

In order to optimize the trade-off in CPSNs, we propose a customizable reliable differential privacy protection model (CRDP). In CRDP, we start from customization of privacy protection levels using differential privacy based on the social distance. Social distance denotes the distance between two users in CPSNs and is represented by least number of hops in this context. Then, a QoS-based mapping function is developed to map social distances to customizable privacy protection levels considering their non-linear correlation. Intuitively, a longer social distance means less intimacy of two users, which demands a higher degree of privacy protection.

In customizable privacy protection cases, the background knowledge attack and collusion attack are additional severe and have always been two primary issues [11]. We formulate them under the framework of ϵ-differential privacy using quantitative measurements. On account of this, we use a modified Laplacian mechanism in which the noise generation process complies with a Markov stochastic process featuring memoryless property. In this way, the correlations among the noises are properly de-correlated, and thereby the composition mechanism no longer unintentionally provide any attack incentives. We also show that the optimized trade-off simultaneously satisfies both customizable privacy protection and the optimality of the Laplacian mechanism. In addition, our extensive experiments demonstrate that the CRDP model has superior performances than current leading models from the perspectives of both customizable privacy protection, high data utility, and attack resistance.

The main contributions of this work are summarized as follows.

- We propose a customizable reliable ϵ-differential privacy model (CRDP) built upon the intimacy of users measured by hop-defined social distance. We devise a QoS-based mapping function which maps social distance to customizable privacy protection levels. Consequently, CRDP achieves flexible privacy protection with less privacy budget while upgrading data utility simultaneously.
- Under customizable privacy protection scenario, we identify background knowledge attack and collusion attack as two leading attacks. The attacks are modelled under the customizable privacy protection framework, followed by quantitative measurement and analysis on them. Built upon this, we reveal the fountainhead of the attacks and then put forward a solution.
- To defeat the two leading attacks, a modified Laplacian mechanism is employed to generate controllable random noises. We further demonstrate that if the noisy generation process complies with a Markov stochastic process, the correlations between the noises can be de-coupled and the composition mechanism is disabled to provide incentive to the attacks. As a result, the CRDP model can minimize the background knowledge attack and eliminate the collusion attack.

4.1.1 Literature Review

Cyber-physical social networks have brought considerable connivence to daily life, but privacy-leaking issues attracts extensive concerns [12]. In CPSN scenarios, there are numerous existing models to provide privacy protection from different perspectives. There are two main branches: data clustering models and differential privacy models [13]. The data clustering model started from K-anonymity [14], and extended to L-diversity [15] and T-closeness [16] to take diversity and distribution into account. The clustering models are practical but limited to scalability [2]. Dwork et al. proposed differential privacy in 2006, which is a pioneering work providing strict privacy protection [17]. In the framework of differential privacy, Laplacian mechanism is widely used to generate real-valued random noisy responses [18]. Differential privacy and its extensions are still fast growing in number [19] and have been applied into practice in various fields, for example, identity privacy [5], pathological archives [20], location privacy [21], mobile devices [7]. The primary issue is that uniform privacy protection is still the mainstream method. All users' privacy protection levels are set as the same despite different requirements of users [22]. This increases overall privacy budget while degrading data utility.

In addition to privacy protection, researchers make considerable efforts on the optimal trade-off. In smart sensing scenario, Wang et al. leveraged Markov decision process establish a model while using reinforcement machine learning to derive the expected trade-off [23]. Similar researches haven been conducted in multi-agent systems [24], distributed algorithms [25], etc. Although there are plentiful existing research on privacy protection and optimization of trade-off, it is not sufficient since the uniform protection method fails to capture the emerging demands of customizable privacy protection. In CPSNs, users with different social distances needs customizable privacy protection levels [5]. A pioneering work of customizable privacy protection discusses the dilemma between conservation or liberty [26]. This is followed by several research of customizable privacy protection, such as data aggregation in CSPNs [27], privacy-preserving anomaly detection [28], and data diffusion over networks [29].

Customized privacy protection can achieve a better trade-off than the traditional uniform privacy protection [30]. Nevertheless, the background knowledge attack and collusion attack become two big issues of customizable privacy [22]. Background knowledge [31] is seldom to be modelled qualitatively since the volume background knowledge is hard to measure. That's also the reason why there is only a potential to minimize it but not fully eliminated. Collusion attack [32] are launched if adversaries believe the collusion can reveals further sensitive information. The incentive of collusion attack in this context is triggered by composition mechanism of differential privacy unintentionally. Therefore, it is possible to fully eliminate it if the incentive is removed by disabling the composition mechanism.

In the case of social distance, there are some existing popular models, for instance, effective distance [33], shortest path distance [34], resistance distance [35], and so on. The proposed model can fit in any distance metrics. To better clarify, we use the shortest path in this section. Other popular distance matrices based privacy protection methods include models proposed by Kasiviswanathan et al. [36] and Xiao et al. [37], which devised a novel scheme to leverage node differential privacy to analyze graph data and developed a differentially private network data sharing model by leveraging structure inference, respectively. However, none of them take customizable privacy protection into consideration.

4.1.2 Customizable Privacy Protection Modelling

In this section, we model the proposed customizable reliable differential privacy protection model (CRDP) built upon the intimacy of users, which is defined by social distance in this section. In cyber-physical social networks (CPSNs), a specific user may share a piece of sensitive data to various recipients while the shared data should be received in different forms according to corresponding privacy demands. The sensitive data in CPSNs are classified as Table 4.3. Intuitively, a longer social distance will result in a higher privacy protection level. The rationale is that sensitive information of users is more confidential to people who are less intimate. In CRDP, we customize the privacy protection levels based on the corresponding privacy requirements rather than the traditional way that fixes the privacy protection level as a constant (Table 4.1).

In CPSNs, the life cycle of the sensitive data starts from being submitted to the service provider. This piece of data is stored in an associated database as a record. If the other users in CPSNs submit a request to access some data including it, the request will be translated to a query to the database. The service provider then processes the data with a randomized differentially private mechanism and responds to the data

Table 4.1 Sensitive data classification in CPSNs

Data classification	Description
Time stamp	$x_i \in R_+$ is positive a real value such as hh:mm:ss
Location	$x_i \in R_3$ is usually the GPS coordinates like longitude
Dualistic states	$x_i \in \{0, 1\}$ represents a binary value like gender or employment situation
Text information	$x_i \in ABC$ represents a serial of alphabets such as address

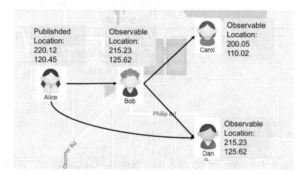

Fig. 4.1 An example of customizable privacy-preserving data sharing based on social distance: Alice is the data curator in this case while Bob and Dan are two one-hop friends and Carol is one two-hop friend. Bob and Dan can observe a more accurate data with less noise comparing to Carol, which is an instance of customization

requestor. We develop a modified Laplacian mechanism to generate noise and deploy it in the server. If another user tries to query the server, the system will compute the shortest social distance and then map it to a customized privacy protection level. Then, the noise is injected to the real output and the noisy output is responded to the requester. The requester is free to conduct any post-processing to realize the value like spreading pattern [38] and recommendation system [39]. In addition, the modified Laplacian mechanism enables an optimized trade-off with reliable resistance against the background knowledge attack and collusion attack.

In Fig. 4.1, the example shows that Alice published her location information which is (220.12, 120.45). For her one-hop friends, Bob and Dan, they can observe the location as (215.23, 125.62). As to her two-hop friend, Carol, her observable location is (200.05, 110.02). Obviously, the friends with longer social distance (hop in this case) obtain the information with more noise. The reason why Bob and Dan observe the same information is that the proposed model leverages the shortest path.

Table 4.2 Notation table

Abbreviation	Description
$\mathcal{M}$	A mechanism complying with DP
LAP	Laplace mechanism of DP
ϵ	Privacy protection level or privacy budget
$S(\cdot)$	Sigmoid function
d	Social distance
θ	Steepness of Sigmoid function
m	The symmetry line of Sigmoid function
k	The mapping coefficient
y	Output with injected noise
DU	Data utility
$Comp(\cdot)$	Composition mechanism of DP
E	Expectation value
v	Value of noise

4.1.2.1 Preliminaries

To better clarify, a notation table and two general concepts are presented in this section, which are the Laplacian mechanism and the composition mechanism in differential privacy (DP) [40]. In privacy-preserving field, differential privacy is the mainstream method because it protects the privacy of strict mathematical guarantee. Laplacian mechanism is a popular randomized mechanism that injects differentially private noise to the original data. Composition mechanism enables the cooperation of different mechanisms so that complex and advanced mechanisms can be developed.

4.1.2.2 Notation Table

In this subsection, we present a notation table that contains most important and global variables to provide guidance of the technical part. The detailed information is as Table 4.2.

4.1.2.3 Laplacian Mechanism and Laplacian Noise

In order to protect privacy in numeral scenarios, Laplacian mechanism injects controllable noises in real-valued scenario to achieve differential privacy.

We use $\{\mathcal{M}: R^n \rightarrow \Delta(R^n)\}$ as the randomized mechanism to inject Laplacian distributed noise N as

$$\mathcal{M}(\mathcal{D}) = \mathcal{D} + LAP\left(\delta\right), \tag{4.1}$$

Fig. 4.2 Composition mechanism with three participants: Alice, Bob, and Carol, have different randomized algorithms with the values of *epsilon* as i, j, and k, the ceiling value of ϵ is $i + j + k$ after the composition mechanism is triggered

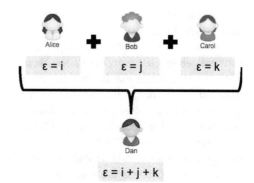

where δ decides the size of noise with global sensitivity and ϵ. Usually, we regard $\mathcal{M}$ as a ϵ-differentially private mechanism if the injected noise follows $LAP\left(\delta\right)$.

4.1.2.4 The Composition Mechanism

The composition mechanism allows various mechanisms to make joint efforts so that complicated and advanced mechanisms can be designed by composing the separated mechanisms as a whole.

Let $\{\mathcal{M}_1, \mathcal{M}_2, \ldots, \mathcal{M}_n : \mathcal{D} \to \Delta(\mathcal{Y})\}$ be the randomized mechanisms which correspondingly satisfy $\{\epsilon_1, \epsilon_2, \ldots, \epsilon_n\}$-differential privacy. Then, we can derive that the mechanism $\{\mathcal{M} : \mathcal{D} \to \Delta(\mathcal{Y}^n)\}$ composed by $\mathcal{M} = \{\mathcal{M}_1, \mathcal{M}_2, \ldots, \mathcal{M}_n\}$ complies with $\sum_i^n \epsilon_i$-differential privacy. The $\sum_i^n \epsilon_i$ is the upper bound of privacy protection level after composition. In this section, we consider $\sum_i^n \epsilon_i$ as the worst case of privacy leakage rooting in composition mechanism.

In Fig. 4.2, we show an example of composition mechanism, where there are three participants, Alice, Bob, and Carol. All three participants have processed the raw data with their associated privacy protection levels: $\epsilon = i$, $\epsilon = j$, and $\epsilon = k$. Dan is the data requestor who can access the differentially private data composed by the three mechanisms, which is denoted by $\epsilon = i + j + k$.

4.1.2.5 Shortest Social Distance Using Dijkstra Algorithm

In CRDP, we use the value of hop to represent the social distance. Usually, there are several paths between two non-adjacent users. We select the shortest path and use the value of its hops as the shortest distance. The reason why we choose the shortest distance is to fairly measure the most intimate relationship of them. To calculate the social distance, we establish a social graph built upon CPSNs using graph theory. Then, the Dijkstra algorithm is used to identify the shortest path. In addition, we pre-set a threshold of maximum distance d_{th} to improve the performance by reducing the computational cost.

The reason why we choose social distance is that it captures the feature of intimacy between two specific users in CPSNs. Therefore, it is suitable to measure relationship of users. In CRDP, we use the shortest path of social graph to represent social distance. It is clear that the distance matrix can be extend to any other ones in various scenarios. In this modelling section, we focus more on the customization and reliability of the privacy-preserving data sharing in CPSNs.

To better analyze a CPSN, we use a social graph $G = \{v_i, e_{ij} | v \in V, e \in E\}$ in this context. A node in G denotes an user in a specific CPSN while a pair of nodes $(u_i, u_j) \in E$ represents the edge, namely, the relationship between the two users. During the whole lifecycle of an user, u_i has several pieces of sensitive data $x_i \in X$, where X is the dataset of sensitive data. The sensitive data x_i will be shared by u_i with a CPSN and be accessed by other users under privacy guarantee. When a different user u_j tries to access the data, the system calculates a noisy output y_{ij} and shares it to u_j based on their social distance d_{ij}. The social distance d_{ij} is a mapping function $\{d_{ij} | V \times V \to R^+\}$ while $\{\epsilon | R^+ \to R^+\}$ that maps the social distances to customizable privacy protection levels $\epsilon\left(\frac{1}{d_{ij}}\right)$. The target of CRDP is to generate approximations $\{y_{ij}\}_{i,j \in V}$ which provides customizable privacy protection while satisfying data utility constraints and reliability requirements. The reliability requires that CRDP can eliminate collusion attacks and background knowledge attacks. To simplify, we use root-mean-square-error (RMSE) to measure the data utility.

We use the classic Dijkstra algorithm to calculate the shortest social distance between two users of CPSN. Firstly, we use the user who shares the data as the source user. Then, the system obtains all the shortest distances from the source users to the other users based on the social graph. In this way, the algorithm can generate a shortest-distance tree with a complexity of $O(|V|^2)$ where $|V|$ is the total number of associated users in a CSPN.

Given u_i as the initial user and social distance d_{ij} be the distance from initial user u_i to target user u_j, the Dijkstra algorithm first assigns initial distances and then improve them progressively. The distance for initial user is set to be 0 while the distances for all the other users are infinity. We give the the initial user u_i a status called "current" and mark all the other users as "unvisited". Based on this, we establish an unvisited set $\{u_j | u_i \in U, j \neq i\}$ which contains all the other users. In the case of the current user, we take all its neighbourhood nodes and compute the tentative distances, respectively. Then, the tentative distances with its current assigned distance are compared to decide which is the minimum value denoting the new current distance. For instance, if current user u_j has a distance of 5 and the edge between A and its neighbour user u_{j+1} has a length of 1, we say the distance through u_j to u_{j+1} is the sum of 5 and 1, namely, 6. In the case that the distance of u_{j+1} was previously set a value larger than 6, the current distance is marked as 6. Otherwise, the current value should be maintained the same.

After all the neighbourhood nodes are traversed, the current user is labelled as "visited" while being deleted from the unvisited set. After all users with greatest threshold are labelled as "visited" or there are infinite smallest tentative distances in the "unvisited set", the current value is regarded as the shortest distance and iteration is suspended. In particular, the left users are supposed to be irrelative to the source user if there are infinite smallest tentative distance.

4.1.2.6 QoS-Based Mapping Function

We use the Sigmoid function $S(\cdot)$ to as the mapping function to calculate customizable privacy protection level ϵ based on social distance d_{ij}. The Sigmoid function $S(\cdot)$ is a popular matrix to measure the satisfactory degree of users regarding quality of service (QoS). We formulate the Sigmoid mapping function as Eq. 4.2.

$$\epsilon_i = S\left(d_{ij}\right) = \frac{k}{1 + \exp(-(d_{ij}) \cdot \theta - m)}, \tag{4.2}$$

where k decides the amplitude of the highest privacy protection level. The parameter θ controls the steepness of middle range of the curve. Besides, m is the arithmetic mean value of $S(\cdot)$.

The reason why we choose Sigmoid function is as follows. We require high customizable privacy protection levels when d_{ij} increases in a low range. Then, after d_{ij} increases across a threshold, the privacy protection level should drop sharply. In addition, further growth of privacy protection level brings marginal benefits when d_{ij} increases in a high range. In Fig. 4.3, we demonstrate the curve transformation with the change of each parameter. Figure 4.3 shows that the Sigmoid function $S(\cdot)$ functions well as the mapping function from the aspect of QoS in CPSNs.

Fig. 4.3 Curve transformation of sigmoid function with different parameter: The parameters, which are k, θ, and m, controls the general trends, which makes it applicable to various scenarios

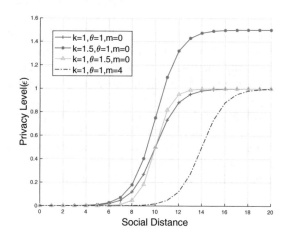

4.1.2.7 Customizable Privacy Protection Model Formulation

Based on the social graph and mapping function in Sects. 4.1.2.5 and 4.1.2.6, we present the model of CRDP which provides customizable privacy protection with attack-proof features in a CPSN. The number of hops is used to measure the social distance and a QoS-based Sigmoid function generates customizable protection levels based on hops. A greater number of hops means a longer social distance and thereby leads to a higher privacy protection level.

A source user u_i shares a piece of sensitive data to the other users $\{u_j | u_j \in U, j \neq i\}$ in a CSPN. The sensitive data should be under privacy guarantee before being shared out. Most current models assume an uniform privacy protection level, which is not feasible, especially in CPSNs, because various relationships exist in it, such as friends, colleagues, playmates, etc. To show the problem clearly, an instance with two customizable privacy protection levels is presented as follows.

Let ϵ_i and ϵ_{i+1} be two different privacy protection levels satisfying $\epsilon_{i+1} > \epsilon_i$, and $M_{\epsilon_i \to \epsilon_{i+1}} : \mathcal{D} \to \Delta(\mathcal{Y}^2)$ be a randomized mechanism. The user u_i publishes two noisy outputs $\{y_{i,j}, y_{i+1,j}\}$, to two other different users, which complies with ϵ_i-DP and ϵ_{i+1}-DP, respectively. We assume these two users are malicious and collude with each other to steal more accurate sensitive information. Then, the privacy protection mechanism should comply with

$$M_{DP}\left(\epsilon_i + \epsilon'_{i+1}\right) = M_{DP}\left(\epsilon_{i+1}\right), \tag{4.3}$$

where ϵ'_{i+1} is the privacy protection level of the second noisy response. The aforementioned composition mechanism shows

$$M_{DP}\left(\epsilon'_{i+1}\right) = M_{DP}\left(\epsilon_{i+1} - \epsilon_i\right), \tag{4.4}$$

where it could be derived that $\epsilon'_{i+1} < \epsilon_{i+1}$. The result means that the privacy protection level of the second user is not relaxed and breaches the assumption $\epsilon_{i+1} > \epsilon_i$, especially when $\epsilon(1) < \epsilon_{i+1} \ll 1$. In this case, the degradation of data causes the infeasibility of the traditional differential privacy mechanism.

With CRDP, we address this problem by customizing privacy protection levels based on social distance using differential privacy. The customizable differential privacy is extended from classic differential privacy as defined below.

Let $\epsilon \geq 0$, $\mathcal{D}$ be the space of the sensitive data, and $\mathcal{A} \subseteq \mathcal{D} \times \mathcal{D}$ to denote an adjacent relation. By taking Eq. 4.2, a randomized mechanism $M \to \Delta(\mathcal{Y})$ is considered to be ϵ-differentially private if and only if

$$\Pr\left[\mathcal{M}(D) \in \Omega\right] = \exp(\epsilon) \cdot \Pr\left[\mathcal{M}(D') \in \Omega\right],$$

$$= \exp\left(k \times \frac{1}{1 + \exp(-\theta \cdot (d_{ij}) - m)}\right) \cdot \Pr\left[\mathcal{M}(D') \in \Omega\right],$$

(4.5)

$$s.t.$$

$$\forall \Omega \subseteq \mathcal{Y},$$

$$\forall (D, D') \in \mathcal{A},$$

where $\mathcal{Y}$ is the noisy outcome. As the protected data is streaming data in this case, the adjacent relationship in this case is defined as the two datasets in adjacent time slots. That means we consider D_t and D_{t+1} as the adjacent datasets. This will enables the dynamic features of the proposed model.

The objective of CRDP is to achieve customizable privacy protection in a CPSN with $\{\mathcal{M} : \mathcal{D} \to \Delta \mathcal{D}_n\}$, which generates n noisy outcomes $\{y_{ij}\}_{j=1}^n$ and $\{y_{ij}\}_{j=1}^n$ and shares the customizable responses to the corresponding users. Therefore, the mechanism $\mathcal{M}$ meets the following constraints.

The first constraint is to provide customizable privacy protection. For each piece of sensitive data x_i, all of its approximations $\{y_{ij}\}_{j=1}^n$ should satisfy $\epsilon\left(\frac{1}{d_{ij}}\right)$-differential privacy based on correspondingly social distance.

The second constraint is decrease the ceiling value of privacy protection level after composition. In terms of all noisy outcomes $\{y_{ij}\}_{j=1}^n$, the maximum value of composed privacy protection is

$$\text{Comp}\left(\epsilon_i\right) = \sum_{i=1, j \neq i}^{n,n} \mathcal{M}_{DP}\left(\epsilon\left(\frac{1}{d_{ij}}\right)\right),$$

(4.6)

where $\mathcal{M}_{DP}$ is the randomized mechanism complying with differential privacy.

The third constraint is that CRDP has the maximum data utility. All noisy outcomes $\{y_{ij}\}_{j=1}^n$ should be as accurate as possible comparing with the actual outcome x_i, which guarantees maximum data utility. In this work, we use root-mean-square-error (RMSE) [41] to measure the data utility, where less RMSE means higher data utility, and vice versa.

In a customizable privacy-preserving data sharing scenario, there are different approximations $\{y_{ij}\}_{j=1}^n$ resulting in multiple values of data utility. Therefore, when we talking about the optimized utility in this work, we mean the sum of data utility as $\sum_{i=1}^n \sum_{j \neq i}^n E\|y_{ij} - x_i\|_2^2$.

Let $\min(DU)$ be the minimum-expected data utility, we formulate the optimized tradeoff as Eq. 4.7.

$$\text{Objective}: \max(DU) \ \& \ \max\left(\sum \epsilon\right)$$

$$s.t.$$

$$\text{customizable } \epsilon_i = \frac{k}{1 + \exp(-(d_{ij}) \cdot \theta - m)} \qquad (4.7)$$

$$\sum M_{DP}\left(\epsilon\left(\frac{1}{d_{ij}}\right)\right) \leq \max M_{DP}\left(\epsilon\left(\frac{1}{d_{ij}}\right)\right),$$

$$\sum E\left\|y_{ij} - x_i\right\|_2^2 \geq \min(DU)$$

Algorithm 1 Customizable Reliable Differential Privacy Algorithm

---

Require: X as the sensitive data of user u_i and d_{th} as maximum social distance;
Ensure: Customizable noisy output Y shared with u_j;
1: Initialize d_t where $d_{ii} = 0$ & $d_{ij} = \infty$;
2: Initialize the source user u_s as *current*;
3: Initialize the status of the other users as *unvisited* and store them in the *unvisited* set;
4: **while** an user u_{ij} within d_{th} is not traversed **do**
5:    Derive the d_ts of u_s's adjacent users;
6:    Compare current value with new d_ts and label the smaller value as the new d_{ij};
7:    Label the status of u_s as *visited*
8:    Delete u_s from *unvisited* set;
9:    Record the shortest social distance d_{ij};
10:    Update $u_c = u_j$ with the shortest d_t;
11: **end while**
12: Another user u_j requests sensitive data X of u_i;
13: Use the Sigmoid function to calculate customizable privacy protection level ϵ;
14: Initialize parameters of the QoS-based mapping function;
15: Implement the modified Laplacian Mechanism to inject noise to X;
16: Return the differentially private response Y

4.1.3 Adversaries and Attacks Modelling

Based on the data sharing scenario modelled in Sect. 4.1.2.7, there are two mainstream types of attacks harassing the customizable privacy-preserving data sharing process in cyber-physical social networks (CPSNs), including collusion attack and background knowledge attack [42]. These two kinds of attacks have already been identified as two primary barriers in privacy field while most other leading attacks are combinations or variants of these two attacks [22]. By mathematical modelling these two attacks, we enable flexible modelling of other leading attacks by simple adjustment. In the customizable privacy-preserving data sharing scenario, we formulate the collusion attack and background knowledge attack under the framework of differential privacy based on their unique features.

4.1.3.1 Collusion Attack

Usually, the definition of collusion attack is two or more adversaries sharing their in-place data with each other to dig out more sensitive data. There are three properties to commit the collusion attack. Firstly, at least two or more adversaries sharing the same interest. Secondly, each of the adversary already has some sensitive data corresponding to the interest in place. Thirdly, the adversaries hold the belief that they can definitely gain more sensitive information after collusion attack.

Under the scenario of Sect. 4.1.2.7, the recipient users $\{j | j \in n, j \neq i\}$ are regarded as adversaries. The customizable privacy protection levels $\left\{ \epsilon\left(\frac{1}{d_{ij}}\right) \middle| \epsilon \in R^+ \right\}$ denote the sensitive data in hands of the adversaries. The composition mechanism is the incentive of adversaries colluding with one another.

Given the noisy responses $\{y_{ij}\}_{j=1}^n$ of x_i from u_i to u_j, each of the $\{y_{ij}\}_{j=1}^n$ submits to $\epsilon\left(\frac{1}{d_{ij}}\right)$-differential privacy. The upper bound of the collusion attack that involves n adversaries is

$$\text{Comp}\left(\frac{1}{d_{ij}}\right) = \sum_{i,j\neq i} M_{DP}\left(\epsilon\left(\frac{1}{d_{ij}}\right)\right), \tag{4.8}$$

where the composition of customizable privacy protection levels equals to the sum of privacy protection levels.

4.1.3.2 Background Knowledge Attack

Background knowledge attack is another fundamental attack in privacy domain. This is natural because an adversary may have some background knowledge, especially in CPSNs, where the users have a relationship with each other. That's also the reason why we believe the users with longer social distance have greater potential threats. There are also three properties for launching a background knowledge attack. First of all, the attack can be launched by only one adversary. Then, an adversary has to have some prior-belief of the interested sensitive data, which is also treated as the background knowledge. Last but not least, an adversary keeps collecting other information to enrich the background knowledge. The attack is successfully carried out when background knowledge is enough.

Under the scenario of Sect. 4.1.2.7, one recipient u_j is regarded as an background knowledge attacker and his prior-belief is assumed as a fixed ϵ_{ad}. Every time the privacy protection level changes from $\epsilon(d_{ij})$ to $\epsilon(d_{ij+1})$, the adversary collects the updated information and composes it with his prior belief until the background knowledge is enough.

Given the multiple released noisy responses y_{ij} of d_{ij} from u_i to u_j, each of the y_{ij} submits to $\epsilon(d_{ij})$-differential privacy. The upper bound of the collusion attack is

$$\text{Comp}\left(s\right) = \mathcal{M}_{DP}\left(\epsilon_{ad} + \sum_{d_{ij}}^{D} \epsilon(d_{ij})\right), \tag{4.9}$$

where the composition of multiple privacy protection levels equals to the sum of privacy protection levels.

4.1.3.3 Universal Attack Modelling

Under the scenario of progressively release sensitive data with customizable differential privacy, we propose the universal attack, which is a generalized form of both collusion attack and background knowledge attack.

There are three reasons that we can formulate these attacks together. Initially, there two attacks are formalized into standard ϵ-differential privacy manner. Sequentially, the attack can be launched because there are more than one $\epsilon(\cdot)$ existing in the attack circumstances. Lastly, the upper bound of composition mechanism provides incentive to the attacks in the same manner.

Given the noisy responses $\{y_{ij}\}_{j=1}^{n}$ of x_i from u_i to u_j, each of the $\{y_{ij}\}_{j=1}^{n}$ submits to $\epsilon\left(\frac{1}{d_{ij}}\right)$-differential privacy. The upper privacy bound of the collusion attack involving n adversaries is

$$\text{Comp}\left(\frac{1}{d_{ij}}\right) = \sum_{i,j \neq} \mathcal{M}_{DP}\left(\epsilon_{ad} + \epsilon\left(\frac{1}{d_{ij}}\right)\right)$$

$$= \mathcal{M}_{DP}\left(\epsilon_{ad} + \epsilon\left(\frac{1}{d_{i1}}\right) + \epsilon\left(\frac{1}{d_{i2}}\right) + \cdots + \epsilon\left(\frac{1}{d_{in}}\right)\right), \tag{4.10}$$

where the composition of customizable privacy protection levels equals to the sum of privacy protection levels.

4.1.4 System Analysis

Based on the system modelling, we show the superiority of CRDP, which satisfies customizable privacy protection, progressively release, and optimized utility. The analysis shows the stochastic process and optimum Laplacian mechanism.

4.1.4.1 Differentially Private Stochastic Process

For n-tuple real-valued data d, we target on developing a differentially private mechanism $\mathcal{M}$ to generate the approximation y_{ij}, in which y_{ij} is sent from u_i to u_j. Two features are required for the mechanism $\mathcal{M}$. Firstly, the accuracy, which is known as data utility in this context, only depends on the social distance $\frac{1}{d_{ij}}$. All other parameters have no negative impact on the data utility. Then, any clusters of CPSN users cannot infer more sensitive information about another user u_i after collusion denoted by $\sum \epsilon(\frac{1}{d_{ij}})$.

Motivated by this, we introduce a private stochastic process defined on a continuous domain. This private stochastic process is used to assist the noise generation and decouple the correlations among noises as follows.

Let ϵ be the privacy protection level, and ϵ_i, ϵ_{i+1}, ϵ_{i+2} ($\epsilon_i < \epsilon_{i+1} < \epsilon_{i+2}$) be three adjacent privacy protection levels, the private stochastic process should possess the following properties.

- The noise complies with Laplacian Mechanism: $\forall \epsilon > 0$, $d\Pr\left(V_\epsilon = v\right) \propto \exp\left(-\epsilon \|v\|_2\right)$;
- The process complies with Markov stochastic process: $\forall \epsilon_i < \epsilon_{i+1} < \epsilon_{i+2}$, $V_{\epsilon_i} | V_{\epsilon_{i+1}}$, $V_{\epsilon_i} \perp V_{\epsilon_{i+2}}$;
- The transfer probability in Markov process is

$$
\begin{aligned}
d\Pr\left(V_{\epsilon_i} = v_i \middle| V_{\epsilon_{i+1}} = v_{i+1}\right) &\propto \delta(v_i - vi + 1) \\
&+ \frac{(n+1)\epsilon_i^{1+\frac{n}{2}} \|v_i - v_{i+1}\|_2^{1-\frac{n}{2}}}{(2\pi)^{\frac{n}{2}}} B_{\frac{n}{2}-1}\left(\epsilon_i \|v_i - v_{i+1}\|_2\right)\tau \\
&+ O\left(\tau^2\right),
\end{aligned}
\tag{4.11}
$$

s.t.

$$
\tau = \frac{\epsilon_i}{\epsilon_{i+1}} - 1,
$$

4.1.4.2 Optimum Laplacian Mechanism

The Laplacian mechanism has already been proved to satisfy ϵ-differential privacy. However, in general cases, the Laplacian mechanism is not optimal as far as minimum mean-squared error. Nevertheless, the Laplacian mechanism should be optimal for the minimum root-mean-square-error (RMSE) if the noise generation process is properly-designed.

Let $\mathcal{M}: R^n \to \Delta(R^n)$ be the ϵ-differentially private mechanism, we have $\mathcal{M}$ which satisfies $y_{ij} = d_{ij} + V$, where $V \sim \rho(V) \in \Delta(R^n)$. The RMSE will be minimal when the designed noise density follows

$$
f_1^n(v) = \left(\frac{\epsilon}{2}\right) \exp\left(-\epsilon \|v\|_1\right),
\tag{4.12}
$$

where $f_1^n(v)$ represents the noise density in the case of v. By substituting Eq. 4.12 to RMSE, we can obtain

$$
\begin{aligned}
E\left\|y_{ij}^t - d_{ij}\right\|_2^2 &= E_{V \sim \rho}\left\|V\right\|^2 \\
&\geq E_{V \sim f_1^n}\left\|V\right\|_2^2 = \frac{2n}{\epsilon^2}.
\end{aligned}
\tag{4.13}
$$

The optimum Laplacian mechanism describes the optimal trade-off in this context. The trade-off is indispensable for the proposed CRDP model as we do need the highest data utility in addition to customizable privacy protection.

4.1.4.3 Mechanism Design and Analysis

To start with, we consider the single-dimension case. The following theorem shows a modified composition mechanism that satisfies all above requirements including customizable privacy protection and universal attack defence.

Two privacy protection levels ϵ_1, ϵ_2, which are abbreviation for $\epsilon_1\left(\frac{1}{d_{ij}}\right)$, $\epsilon_2\left(\frac{1}{d_{ij}}\right)$. They correspond to two random mechanisms: $\mathcal{M}_1$ and $\mathcal{M}_2$. We assume $0 < \epsilon_1\left(\frac{1}{d_{ij}}\right) < \epsilon_2\left(\frac{1}{d_{ij}}\right)$, which is also workable otherwise. Based on this, the mechanism can be represented as

$$
y_{i1} = d + V_1, \; y_{i2} = d + V_2, \; (V_1, V_2) \sim \rho\Delta(R^2).
\tag{4.14}
$$

Moreover, the density $f_{\epsilon_1(\frac{1}{d_{ij}}), \epsilon_2(\frac{1}{d_{ij}})}$ is

$$
\begin{aligned}
f_{\epsilon_1, \epsilon_2}(x, y) &= \frac{\epsilon_1^2}{2\epsilon_2} \exp\left(-\epsilon_2|y|\right)\delta(x - y) \\
&+ \frac{\epsilon_1(\epsilon_2^2 - \epsilon_1^2)}{4\epsilon_2} \exp\left(-\epsilon_1|x - y| - \epsilon_2|y|\right).
\end{aligned}
\tag{4.15}
$$

With above analysis, we summarize the following properties of the above theorem.

- The random mechanism $\mathcal{M}_1$ is $\epsilon_1\left(\frac{1}{d_{ij}}\right)$-differentially private;
- The random mechanism $\mathcal{M}_1$ is optimal. Namely, $\mathcal{M}_1$ minimizes the mean-squared error $E(V_1)^2$;
- The random mechanism $\mathcal{M}_2$ is $\epsilon_2\left(\frac{1}{d_{ij}}\right)$-differentially private;
- The random mechanism $\mathcal{M}_2$ is optimal. In particular, $\mathcal{M}_2$ minimizes the mean-squared error $E(V_2)^2$;
- The random mechanism $(\mathcal{M}_1, \mathcal{M}_2)$ is $\epsilon_2\left(\frac{1}{d_{ij}}\right)$-differentially private.

4.1.5 Performance Evaluation

To show the performances of the proposed model, we begin with testifying the trade-off of CRDP. Followed by this, we further demonstrate the performances of CRDP over classic ϵ-differential privacy (CDP) and the classic customizable differential privacy (CCDP) against background knowledge attack and collusion attack. The obtained experimental results on real-world datasets are superior comparing with the two baseline models and thereby confirm the superiority of CRDP.

We evaluate our model on the "Google +" dataset collected by *Jure Leskovec* [43]. This dataset contains $107, 614$ Nodes and $13, 673, 453$ edges. We use the Dijkstra algorithm to find the shortest social distance and use it to customize the privacy protection levels. This can be extended to any other kinds of distance metrics. The algorithms are implemented on Matlab 2017a and are executed on Mac OS platform with Core I5@2.7 GHz CPU and $8G$ memory.

In the experiments, we compare the proposed customizable reliable ϵ-differential privacy (CRDP) model with both classic ϵ-differential privacy (CDP) and classic customizable ϵ-differential privacy (CCDP). In classic ϵ-differential privacy, the privacy protection level is fixed for all the circumstances and the generated noises complies to the laplacian mechanism. While in classic customizable ϵ-differential privacy, the privacy protection level is customizable and the noises complies to the laplacian mechanism. In the proposed customizable ϵ-differential privacy model, the privacy protection level is customizable and the noises complies with the proposed theorem. The parameter initialization is discussed in following subsections.

In the following experiments, we assume the maximum social distance is 6 based on the famous theory entitled Six Degree of Separation [44–46]. We randomly partition 1 piece of the above CPSN with $1, 500$ nodes. In the segmented network, we identify there are only 4 users with a maximum social distance of 6 while none of the users have a social distance over 6, which also testifies the assumption. We pick the one at the middles of this network.

4.1.5.1 Privacy Protection Levels

In term of privacy protection levels, we consider the both single CPSN and multiple CPSNs case. In single CPSN, the privacy protection level is fixed and the data is released once and for all, therefore all the other users can access the same data with the same privacy protection level. However, although the privacy privacy protection level is fixed in different privacy networks, the randomness of Laplacian mechanism lead to different noisy responses. Thus, users may access different data with the same privacy protection level. The differences are clearly shown in Fig. 4.4.

In Fig. 4.5, the privacy protection level of CDP remains 2.63 while the privacy protection levels of CCDP and CRDP increase from 5.36 and 0.15 to 10.88 and 2.63, respectively. Between the two models, the privacy protection level of CCDP increases with the increment of the privacy protection levels' sum while the maximum privacy

Fig. 4.4 Data utility
comparison among three
models: The measurement
matrix is MSE-based,
therefore, the utility upgrades
with the decrease of the
value. Therefore, CRDP has
the best performance
regarding data utility

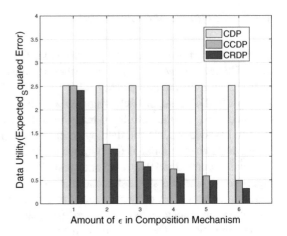

protection level of CRDP equals to the largest privacy protection level. Based on the
results, we can tell the CDP is not affected by composition mechanism but maintains
a relative high privacy protection level, which is 2.63 throughout. The CCDP suffers
from composition mechanism and the privacy protection keeps degrading, which
makes it unacceptable. In the case of CRDP, the privacy protection level keeps in
a low and steady situation and the maximum value equals the the largest existing
privacy protection level. The amplitude increase is 2.48, which confirms high-quality
customizable privacy protection of CRDP.

In Fig. 4.5, the privacy protection levels of all three models increases with the
increment of ϵ quantity. CDP increases in a linearly style while CCDP and CRDP
increase in an exponential style. In this case, the only difference is that the CDP
is affected by composition mechanism and the privacy protection level increases
rapidly. As the privacy protection levels of CDP maintain 2.63, the privacy protection
level of CDP increases lineally to 6×2.63 in the end. CRDP has a higher privacy
protection level by 45.5% and 83.2% compared to CCDP and CDP. In this case, the
CRDP embodies better performances than CDP and CCDP.

4.1.5.2 Data Utility

To begin with, we use a random algorithm to choose 6 customizable privacy protection levels from the interval of $\epsilon \in [0, \infty)$ based on social distances. Then the
fix privacy protection level is chosen to be the minimum one. Laplacian mechanism
is performed to generate noisy responses. At last, the composition mechanism is
performed. The data utility are calculated by RMSE and Eq. 4.13.

In Fig. 4.5, we can see the expected-squared error of CDP maintains as 2.52 and
the value is the highest throughout. Although the expected squared errors of CCDP
and CRDP keep decreasing, the CRDP enjoys higher descent amplitude from 2.46 to
0.33. This illustrates that CDP has relative lower data utility compared to CCDP and

Fig. 4.5 Privacy protection level comparison in term of customizable ϵ based on social distance: **a** In the scenario of various users with different social distances, the CRDP provides flexible privacy protection while maintaining high protection level and data utility. **b** In a potential scenario when the users are in different CPSNs, the superiority of CRDP is more significant

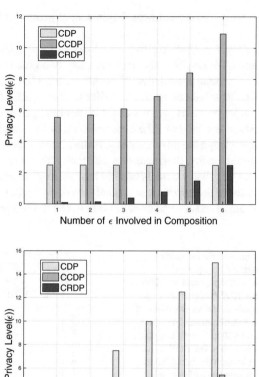

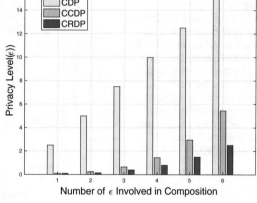

CRDP while CRDP enjoys the optimal data utility among all of the three models. The histogram shows the data utility of CRDP increases by 8% overall in comparison to CCDP, not to mention CDP. The results conform to the above analysis.

4.1.5.3 Performances Against Background Knowledge Attack and Collusion Attack

We focus on the performances against background knowledge attack and collusion attack in this subsection. The two types of attacks are discussed correspondingly.

4.1.5.4   Performances Against Background Attack

In Fig. 4.6, there are two stair-lines and a straight-line representing CCDP, CRDP, and CDP, respectively. There are also two green dashed-lines to denote the possible background knowledge of an adversary. CDP keeps a constant while CRDP and CCDP increases as the stair-lines. The smaller the ϵ is, the better protection the model provides. As we use ϵ-differential privacy to model the prior belief of the adversary, we have the observation that the smaller the ϵ is, the more background knowledge the adversary has. If the bottom green dashed-line ($\epsilon = 0.47$) represents the adversary, that means the adversary nearly knows all the background knowledge and all privacy models can not provide any protection. If the adversary's prior belief is above the upper green dashed-line ($\epsilon = 2.85$), the CDP and CRDP fully functions. If the adversary's prior belief is between the two green dashed-lines, that means only

Fig. 4.6 Performances against background knowledge attack in CPSNs: **a** CRDP can guarantee the background knowledge will not enrich even if multiple users share their own data to each other. **b** CRDP can also guarantee the background knowledge will not enrich even if multiple users share their own data across multiple CPSNs

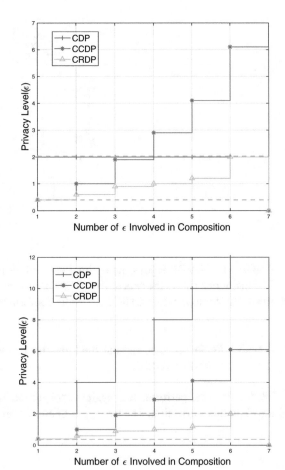

the proposed CRDP can protect the sensitive information. Therefore, CRDP is the best way to provide privacy protection in CPSNs.

In Fig. 4.6, the CDP increases by a large amplitude to nearly 12. Thus, it can not provide any protection in term of background knowledge attack. Although the CCDP can provide partial protection, it fails to function after $\epsilon \geq 4$. But the CRDP still functions well in the range between the green dashed-lines from $\epsilon = 0.47$ until $\epsilon = 2.85$.

All in all, we can conclude the CRDP is the only model which can fully function in both occasions. CDP and CCDP suffers from background knowledge attack with varying degrees in different scenarios.

4.1.5.5 Performances Against Collusion Attack

Collusion attack appears under the CPSN scenario. The differences between collusion attack and background attack are discussed in Sect. 4.1.3. In the case of defence, collusion attack can be eliminated in the proposed model. With CRDP, the incentive of collusion is dispelled so that the collusion attack can not be launched.

In Fig. 4.7, privacy protection level of CDP maintains a constant 0.47 while CRDP and CCDP increases as the fold-lines. As the cases in Sect. 4.1.5.1, we can see the CDP is still free of composition mechanism, which means CDP is unaffected regarding to collusion attack. However, privacy protection levels of CRDP and CCDP increases with the increment of ϵ's quantity from 0.47 and 0.47 to 2.05 and 6.91, respectively. Despite the similar increasing trend, there is a big difference. The CCDP suffers from collusion attack as two or more adversaries can gain more information from the sum of ϵs. The advantage of CRDP is that two or more adversaries gains no further information after collusion. The reason is that the sum of ϵs equals to the maximum ϵ after collusion. Therefore, the adversary with lowest ϵ, namely, 0.47 has no incentive to collude with others. CRDP thereby chops off the root of collusion attack.

In Fig. 4.7, the only difference is that performance against collusion of CDP degrades severely. Collusion attack can easily breach the privacy of CDP in multiple CPSNs as shown. The sum of ϵs increases linearly until 6×0.47. As for CCDP and CRDP, the circumstances remains the same. We can easily conclude that CRDP functions well in multiple CPSNs case.

To conclude, the proposed CRDP shows the higher performances in term of eliminating the incentive of collusion attacks.

4.1.5.6 Performance of Cost

To measure the costs of CRDP and two baseline models, we choose to use the processing time as the index. We measure the time consumption of all three models against the increment of data size by 250 when ϵ is set to be 3. The shortest distance of all users is known and stored by service providers as part of the network statistics. Therefore, we will not consider the processing time of the calculation of shortest

Fig. 4.7 Performances
against collusion attack in
CPSNs: In both figures, the
collusion attack could be
eliminated because the
collusion can only help to
improve the data accuracy of
users with less accurate data.
For the users with
high-accurate data, there is
no incentive for them

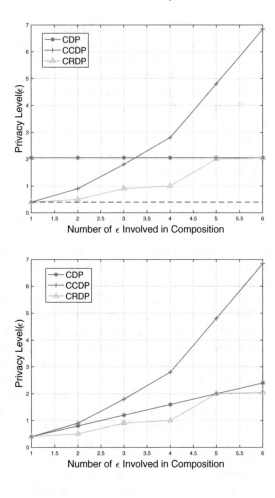

distances in this scenario. The processing time in Fig. 4.8 shows that all processing time is linear to the volume of data size. CRDP and CCDP have similar processing time in average. They have a relatively longer processing time compared with CDP and the overflow rate is about 15%. The randomness of processing time of CRDP and CCDP is resulted by the highest value of the social distance in the data sample, usually ranges from 4 to 6. The processing time is in *ms* level while the deployment of CRDP causes only 15% of extra time consumption compared with CDP and 0% compared with CCDP. Consequently, the performance of cost is acceptable considering the realization of flexible privacy protection and optimized data utility.

Fig. 4.8 Processing time comparison of three models: CRDP and CCDP have similar processing time in average. They have a relatively longer processing time compared with CDP but the overflow rate is about 15%

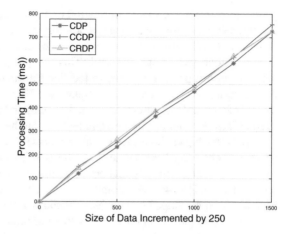

4.1.6 Summary

In this work, we show the disadvantage of fixed privacy protection levels and the major flaws of current customizable privacy protection methods. Our analysis shows that there is either privacy leakage or over-protection if privacy protection level is fixed for all parties. To address this, we develop a novel customizable reliable privacy protection (CRDP) model on top of differential privacy, in which the privacy protection level is customized by social distance. We calculate social distance using the shortest path algorithm and pre-set a threshold to reduce the computational cost, which is followed by a QoS-based mapping function which maps social distance to customizable privacy protection levels. Built upon this, we further develop an advanced mechanism to sample the Laplacian noise complying with a Markov stochastic process. The correlations among noises are then de-coupled and composition mechanism cannot provide any incentive to attacks. With the new mechanism, we derive the optimized trade-off while the background attack is minimized and the collusion attack is eliminated. Extensive experiments are implemented to show the superiority of CRDP over existing works from the perspectives of customizable privacy protection and attack resistance.

4.2 Personalized Privacy in Social Networks Using Differential Privacy

Social networks have become an important part of human society. According to the latest statistics [47], Facebook has had more than 2.7 billion active users in 2020 while it had only 100 million in 2008. This rapid and continuous growth of social networks indicates that communication on them has become a prominent method for people to connect and share information on the Internet. Furthermore, people even use these services for their business promotion, such as advertising and marketing activities. Social networks have become more ubiquitous due to the new advances in smartphone technology [48, 49]. This has provided an opportunity for social network service providers to utilize location information of users in their services. For example, Facebook Places, Foursquare and Yelp are popular social networks that mostly rely on utilizing users' location data in their services. They offer a variety of useful services, from location recommendation to nearby friend alert [50].

However, a big challenge for social networks is how to protect location privacy of users. This challenge has become one of the most important issues in social media due to the existing structure of social networks that enables an adversary to track movements of users [48, 49]. For example, a new Chrome extension called Marauder's Map has been developed that enables Facebook users to easily track movements of other users and plot them on a map with an accuracy of around one meter [51]. It uses the location data that users have shared in Facebook Messenger chats. Moreover, different methods have been proposed for user location inference based on users' tweets [52–54]. This is really a big issue since other private information of users can be revealed by analyzing their location data (e.g., home address, health condition, interests, etc.).

To address privacy issues, social network service providers offer some built-in tools enabling users to decide on their own privacy preferences. In addition, different methods have been proposed to protect user location privacy in social networks and Geo-Social Networks (GeoSNs) [55–58]. However, these tools and methods introduce additional problems that may lead to further privacy leakage as follows.

Firstly, current solutions rely on user collaboration while some users may not be competent enough to collaborate in such processes. Moreover, some users are not even aware that social networks have been equipped with these privacy protection tools. They might customize their default privacy settings only after their privacy is violated [51, 59, 60]. Secondly, the mentioned privacy protection tools and methods are not efficient enough to protect different users' privacy requirements [51, 55, 60]. Specifically, they are not flexible in terms of social distance between users and rigidly divide users to be either friends or strangers [61]. These privacy protection tools look at the level of privacy protection as a rigid binary function, while in reality, we treat privacy differently against different relationships. Although differential privacy is the dominant tool used for privacy protection, it cannot offer customized privacy protection in its current form. Finally, applying rigid privacy policies keeps users information local and limits data utility for public [61].

To address the aforementioned problems, we propose a Distance-Based Location Privacy Protection (DBLP2) mechanism in this section. The proposed mechanism protects location privacy of social network users based on their social distances. We define social distance as a measurement index of social relationship which indicates the intimacy of users based on their interactions in the social network. To offer customizable privacy protection, we extend the standard differential privacy framework. For this purpose, we add variable artificial noise to the real location of a user in order to obtain a sanitized location. The amount of noise is decided based on the social distance between an information requester and the information provider. In the proposed mechanism, a smaller social distance indicates a closer relationship, as a result, less noise is added to the location data. Since the proposed location privacy protection process is run automatically by the system, it does not rely on users' collaboration. In addition, the proposed mechanism keeps a balance between data utility and privacy protection by generating responses with optimal accuracy. Consequently, it improves utility of the whole network.

We conducted an extensive security analysis on the proposed system. The results show that it is resilient to post processings, i.e., performing computation on the system output cannot weaken its privacy guarantees. Furthermore, we prove that the system is immune to collusion attacks, in which a group of users collaborate and share their responses to make a better approximation of a user's private location. Hence, the result of any collusion attack gains no more extra information. As a result, no additional privacy-preserving noise is required when multiple users ask for the location of the same user.

Our main contributions are listed as follows.

- We extend the traditional differential privacy framework to customizable differential privacy. The proposed scheme can offer privacy protection at individual level, which is desired by various users. To the best of our knowledge, this is an early work in personalized privacy protection.
- We propose a distance-based and customizable location privacy protection mechanism DBLP2 for social network users by extending the standard differential privacy framework. The proposed mechanism provides a flexible location privacy protection framework without requiring users' collaboration. In addition, it improves data utility by providing privacy-aware access rights and generating responses with optimal accuracy.
- We develop a weighted and directed graph model to measure the social distance between users by customizing the concept of effective distance.

4.2.1   Literature Review

Privacy protection in social networks have been studied comprehensively. Abawajy et al. [59] have analyzed different privacy risks and attacks in social media along with the presentation of a threat model. They have also quantified and classified the background knowledge which is used by adversaries to violate users' privacy. In addition, Fire, Goldschmidt and Elovici [60] presented some strategies and methods in privacy-preserving social network data publishing through a detailed review of different security and privacy issues. They have reviewed a range of existing solutions for these privacy issues along with eight simple-to-implement recommendations which can improve users' security and privacy when using these platforms [60].

A few location privacy protection mechanisms have been proposed based on differential privacy. A perturbation technique based on differential privacy was introduced [62] to achieve geo-indistinguishability for protecting the exact location of a user. This technique adds random Laplace-distributed noise to users' location in order to sanitize their location before publishing. A differentially private hierarchical location sanitization (DPHLS) approach has been proposed for location privacy protection in large-scale user trajectories. The approach provides a personalized hierarchical mechanism that protects a user's location privacy by hiding the location in a dataset which includes a subset of all possible locations that might be visited in a region [56]. By doing this, the level of location randomization is reduced, hence, the amount of noise required for satisfying differential privacy conditions is minimized.

Another research study in the differential privacy field has been conducted [63] to consider the problem of releasing private data under differential privacy when the privacy level is subject to change over time. In spite of other works that consider privacy level as a fixed value, they have studied cases in which users may wish to relax their privacy level for subsequent releases of the same data after either a re-evaluation of the privacy concerns or the need for better accuracy. For this reason, the authors have presented a mechanism whose outputs can be described by a lazy Markov stochastic process to analyze the case of gradual release of private data.

Some other research studies have recently been done on the location privacy of Geo-Social Networks (GeoSNs) users [55, 57, 58]. GeoSNs are a variety of social networks by which users can find their favorite events, persons or groups in a specific region or identify popular places by comparing how many people have already checked-in at different places. This is done by utilizing users' location data which have been shared by them in that region. In fact, GeoSNs combine location recommendation services (such as services offered by location-based services) with social network functionality [64, 65]. In other words, they can be viewed as location-based social networks which connect people in a specific region based on their interests.

In [65] different GeoSNs were classified into three categories *Content-Centric*, *Check-In Based* and *Tracking-Based* according to the services they offer. In addition, the main privacy issues that threaten user location privacy were identified. Moreover, the authors of [57] have studied techniques that sanitize users' location data based on differential privacy framework before publishing them as location recommendations in GeoSNs. Moreover, to enhance the accuracy of the location recommendations, they have identified some effective factors which improve data accuracy.

In [58], a location-privacy-aware framework is offered to publish reviews for local business service systems. The proposed framework publishes reviews based on utility to achieve two main goals, maximizing the amount of public reviews which users share and having the maximum number of businesses that obey the proposed public principle. Moreover, in [66], the differential privacy framework has been adopted to the context of location-based services to quantify the level of indistinguishability in the users' location data. Their proposed scheme is a symmetric mechanism that injects noise to the real location of the user through a noise function to obfuscate the user's location before its submission. They have also analyzed the mechanism with respect to location privacy and utility.

One of the latest research on the location privacy of GeoSNs users is [67] in which the importance of users' awareness of the outcomes of sharing their locations in GeoSNs along with the resultant privacy threats were discussed. Moreover, a feedback tool has been designed to enable users to realize the level of threat related to the disclosure of their location data. To evaluate the effectiveness of the proposed feedback tool, they have conducted a user study which confirms the necessity of users' location privacy awareness.

4.2.2 Preliminaries

This section presents the foundation for the next sections. After briefly reviewing the concept of differential privacy and the Laplace mechanism, we introduce the necessity of customizing the adjacency relation defined in the standard differential privacy to match its definition with the location domain.

4.2.2.1 Differential Privacy

Differential privacy is a privacy preserving framework that enables data analyzing bodies to promise privacy guarantees to individuals who share their personal information. In fact, differentially private mechanisms can make users' private data available for data analysis, without needing data clean rooms, data usage agreements or data protection plans. More precisely, a differentially private mechanism that publishes users' private data provides a form of indistinguishability between every two adjacent databases. Here, "adjacent" means that they differ only in a single record. However, as you see later, we will extend the concept of "adjacency" to the location domain.

Differential Privacy [17]: The randomized mechanism $\mathcal{A}$ with domain H is ϵ−differential private if for all $S \subseteq Range(\mathcal{A})$ and for all adjacent $x, y \in H$ (i.e. $||x - y||_1 \leq 1$) we have

$$Pr[\mathcal{A}(x) \subseteq S] \leq e^\epsilon Pr[\mathcal{A}(y) \subseteq S],$$

where ϵ is the privacy level which is a positive value and denotes the level of privacy guarantees such that a smaller value of ϵ represents a stricter privacy requirement. In other words, for a smaller ϵ, the mechanism makes any adjacent data x and y more indistinguishable, i.e. for a small value of ϵ, with almost the same probability, the published $\mathcal{A}(x)$ and $\mathcal{A}(y)$ are placed in the same region S. However, for a large ϵ, this probability is much higher for $\mathcal{A}(x)$ than $\mathcal{A}(y)$ which makes them more distinguishable.

Therefore, mechanism $\mathcal{A}$ can address privacy concerns that individuals might have about the release of their private information. Note that differential privacy is a definition, not an algorithm. In other words, we can have many differentially private algorithms for a privacy scenario and a given ϵ.

4.2.2.2   Laplace Mechanism

One of the most popular mechanisms developed based on the differential privacy framework is the Laplace mechanism [68, 69] in which Laplace-distributed noise is added to users' private data to make it ϵ−differentially private.

Laplace Mechanism [69]: Given the private data $x \in H$, the Laplace mechanism is defined as:

$$\mathcal{A}_L(x, \epsilon) = x + N, \tag{4.16}$$

where, N is Laplace-distributed noise with scale parameter $1/\epsilon$ and zero mean, i.e.,

$$N \sim Lap\left(0, \frac{1}{\epsilon}\right) \tag{4.17}$$

The probability density function for N is:

$$f_N(n) = \frac{\epsilon}{2} e^{(-\epsilon|n|)}, \tag{4.18}$$

where ϵ denotes the privacy level required by the user. The Laplace distribution is a symmetric version of the exponential distribution. According to its probability density function, with high probability, the Laplace mechanism generates much stronger noise for small values of privacy level and vice versa [68].

In this section, we consider the set of private data $H \subseteq R^2$ since our target is to protect users' location data which is assumed as $L = < latitude, longitude >$ where $latitude, longitude \in R$ are GPS coordinates in the ranges $[-90, 90]$ and $[-180, 180]$ respectively. Moreover, the adjacency relation defined in the standard differential privacy should be customized, since we need a mechanism for publishing location data which guarantees that *adjacent* locations are indistinguishable to some extent. For this reason, we will customize the adjacency relation definition later in Sect. 4.4 in order to use differential privacy framework in the location domain.

4.2.3   *The Proposed DBLP2 Mechanism*

In this section, the proposed DBLP2 mechanism is presented. Firstly, we present the system architecture and propose a graph model for social networks. Then, we discuss how social distances are converted to privacy levels. Finally, we present the proposed customizable differential privacy framework. The designed mechanism is independent of user collaboration and improves the utility of social networks. In other words, it satisfies the following properties:

Flexible privacy: The system must generate $\epsilon(d_{ij})$-differential private responses, where d_{ij} is the social distance between user u_i and u_j. Thus, the privacy level ϵ must be a function of social distances.

Independent of user collaboration: The system must embrace the whole responsibility of users' privacy protection regardless of whether users collaborate with the system or not. Therefore, by default, the system must perform a standard distance-to-privacy function for each user to obtain the required privacy levels against other users. Competent users can customize this function based on their own requirements.

Optimal accuracy: Responses generated by the system must be as accurate as possible regarding the trade-off between privacy protection and data utility. Therefore, to preserve data utility, the level of location generalization must be kept to a minimum, i.e., the system needs to minimize the expected squared error $\| L_{ij} - L_i \|^2$ $(i, j \in V)$, where L_i is the real location of user u_i and L_{ij} is an approximation of L_i generated by the system for sending to user u_j.

4.2.3.1 System Architecture

In this work, we assume the social network service provider as a centralized trusted entity that is in charge of keeping users' raw private location data, calculating the social distances and executing our proposed DBLP2 mechanism.

Figure 4.9 shows the proposed system architecture. As you see, when Bob sends a request for Alice's Location data, firstly, using their social distance, i.e. $d_{Alice,Bob}$, the privacy level ϵ that Alice requires against Bob is obtained. The required privacy level ϵ is calculated by a distance-to-privacy function f. Default or Alice settings have a critical role to convert $d_{Alice,Bob}$ to ϵ. Since function f can be different for different users (depending how location privacy is important for the user), it must be customizable by users based on their requirement. The default settings are designed

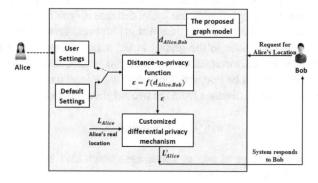

Fig. 4.9 The proposed DBLP2 system architecture

based on the behavior of non-competent users. As you will see in Sect. 4.3, these default settings model a moderate behavior which most users have in social networks in terms of privacy protection. Obviously, Alice can personalize these settings based on her privacy protection requirements.

Finally, using a customized differential privacy mechanism, an appropriate amount of noise (regarding the obtained privacy level ϵ) is injected to Alice's real location (L_{Alice}) and the sanitized location L'_{Alice} is generated for offering to Bob. In the next subsections, details of the mentioned stages are discussed.

4.2.3.2 Graph Model

We model social networks by a directed and weighted graph $G = (V, E)$ in which nodes represent social network users and edges define social relations between users. Therefore, if user u_i has $|u_i|$ friends in her friend list, node u_i is connected to a set of $|u_i|$ neighboring nodes. Now suppose the graph has $|V| = N$ nodes and for each edge $(i, j) \in E$ we assign a weight w_{ij} which represents the social distance between user u_i and u_j. In most cases, a social network user has different social distances from other users who are in her friend list. For example, although a family member and a colleague of her can both be in her friend list, she is more comfortable with the family member than the colleague in terms of privacy concerns. Hence, we believe that weighted graphs are more appropriate models for social networks rather than unweighted graphs because they enable us to model different social distances by weighted edges.

Moreover, we adopt a directed graph to model the network instead of undirected because we believe that social relations between users are not necessarily symmetric. In other words, two friends in a social network might have different feelings about each other. For example, although Bob regards himself as very close to Alice, she may consider some privacy protection settings against Bob. We call this attribute *friendship asymmetry* in social networks. A directed graph model allows us to analyze privacy protection requirements for each user separately. Therefore, for any given users u_i and u_j, equations $d_{ij} = d_{ji}$ and $w_{ij} = w_{ji}$ are not necessarily true.

Using the proposed graph model, the social distance d_{ij} can be obtained. For this reason, we extend the effective distance definition [70] to obtain the distance between friend users (neighbor nodes in the graph) in the social network (or equivalently w_{ij} where $(i, j) \in E$). However, other methods and techniques for social distance measurement [71, 72] can be integrated into the DBLP2 mechanism.

The extended effective distance from the two neighbor nodes u_i and u_j is defined as

$$e(i, j) = 1 - log(p_{ij}), \qquad (4.19)$$

where p_{ij} is the percentage of user u_i's messages which have been sent to user u_j, (i.e., $0 < p_{ij} \leq 1$) and is calculated by

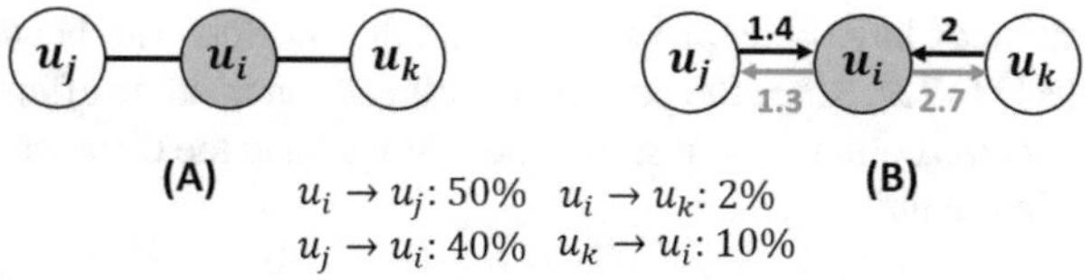

(A) $u_i \to u_j$: 50% $\quad u_i \to u_k$: 2% **(B)**
$u_j \to u_i$: 40% $\quad u_k \to u_i$: 10%

Fig. 4.10 A simple example showing three users of a social network modeled by a simple graph. **a** In the current privacy protection schemes adopted by social network service providers, all of a user's friends have the same access rights regardless of how frequent they have been in contact with that user. Moreover, the friendship asymmetry attribute is not considered in these models. **b** Since most of user u_i's messages (50%) have been sent to u_j, after applying effective weights to the graph, u_j has a smaller distances to u_i than u_k. The distance between two non-neighbor nodes can be obtained by adding the individual weights of the shortest path between them

$$p_{ij} = \frac{m_{ij}}{\sum_{k=1}^{|u_i|} m_{ik}}, \tag{4.20}$$

where m_{ij} is the number of messages that user u_i has sent to user u_j and $|u_i|$ is the cardinality of user u_i (the number of u_i's friends).

The concept of effective distance reflects the idea that a small value of p_{ij} or equivalently a small number of messages exchanged between user u_i and u_j results in a large distance between them, and vice versa. Therefore, for each edge $(i, j) \in E$ we adopt

$$w_{ij} = e(i, j), \tag{4.21}$$

as the effective weight that represents the social distance between two friend users u_i and u_j.

A simple example is illustrated in Fig. 4.10 which shows how effective weights are applied to the nodes of a social network graph. As you see, 50% of user u_i's messages has been sent to u_j (i.e., $P_{ij} = 0.5$) while she has sent only 2% of her messages to u_k. Therefore, after calculating effective weights for each friend and applying them to the graph. You see that u_i has a smaller distance to u_j than u_k. The friendship asymmetry attribute is also considered in our model which makes the social network graph a directed graph.

By applying effective weights to the whole network's graph, we are able to calculate the distance between non-friend users. For this reason, we just need to add individual effective weights on each path between two non-neighbor nodes and find the path with the minimum additive effective weights, i.e.

$$d_{ij} = min \left(\sum_{l=1}^{K_p} w_l^p \right), \tag{4.22}$$

where w_l^p is the effective weight of the lth edge on the pth path between node u_i and u_j and K_p is the the number of edges that make path p.

Different methods have been proposed to find the shortest path between a pair of nodes in graphs [71, 72]. Since the purpose of this section is not to offer an algorithm for the shortest path problem, we just assume that we have the distance between any pairs of nodes in the network.

4.2.3.3   Converting Social Distances to Privacy Levels

Before injecting noise to a user's location data, we need to quantify her privacy level against other users in a social network since we need to design a system with flexible (variable) privacy level. Hence, we adopt the social distance as a determinant factor to obtain different privacy levels that a user requires against other users.

To discuss how social distances are mapped to privacy levels, we assume f is a function which converts social distance between user u_i and u_j, i.e. $d_{ij}(i, j \in V)$, to a privacy level $\epsilon(d_{ij})$. The following properties can be considered for a standard function f in social networks.

- f is a decreasing function since the standard differential privacy definition specifies that a larger value of ϵ represents a more relaxed privacy level (or equivalently a small social distance) and vice versa. Thus, there is always an inverse relationship between d_{ij} and privacy level ϵ. The slope of these inverse changes depends on the user behavior in terms of privacy protection, thus, it can be different for each user.
- For large distances ($d \to \infty$), ϵ must be near zero ($\epsilon \to 0$). This means a tight privacy constraint for strangers who are far from a user in the network.
- For small distances, i.e. $d \to 0$, ϵ must be a relatively large value ($\epsilon \gg 1$) which represents a loose privacy constraint for a user's close friends in the network.

Different functions can be defined with the mentioned properties. For example, an exponential function f in the following can be adopted to convert social distances to privacy levels.

$$f(d_{ij}) = e^{(a - b d_{ij})} , \qquad (4.23)$$

where, $a, b > 0$ are regression coefficients used to calibrate the formula. However, function f can have different properties for different users (dependant on how privacy is important for each user). For example, a user might be very conservative and only allows her family members and close friends to see her location. On the other hand, there are always some social network users with minimal privacy concerns. Hence, a single function f can not satisfy privacy requirements of all users with different privacy protection requirements. Therefore, users should be able to customize f based on their own requirements.

However, to make the system independent of user collaboration, we consider the behavior of the moderate user shown in Fig. 4.11 as a standard model and adopt its function as the standard function f for all users. Those users who want to customize this function can change the related settings. For example, by applying constants c_1, c_2 and c_3 to the mentioned function f, we obtain the following function f'.

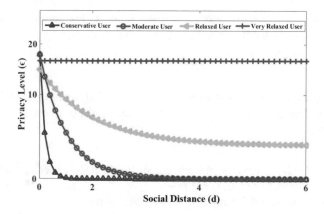

Fig. 4.11 An example of four users with different privacy protection requirements. The conservative user is relaxed only with her family members and close friends (small social distances) and needs strict privacy protection (small ϵ) as distance is increased. On the other hand, the very relaxed user has no concern about his privacy in the network

$$\epsilon = f'(d_{ij}) = c_1 + e^{(c_3 - c_2 d_{ij})}, \tag{4.24}$$

where $c_1, c_2, c_3 \geq 0$. A default value is defined by the system for constants c_1 to c_3 to create the standard function. However, each user is able to customize the function by changing the appropriate settings. Therefore, all four groups of users introduced in Fig. 4.11 are covered through a single function.

4.2.3.4 Customizable Differential Privacy

After discussing how social distances are converted to appropriate privacy levels, we are ready now to present the noise injection mechanism for the DBLP2 system. We adopt the differential privacy framework because of its verified privacy guarantees. The target is to randomize a user's real location such that there must always be a minimum level of indistinguishability for an adversary between the user's real location and any other location which is adjacent to it. This level of indistinguishability is varied inversely with the privacy level ϵ, i.e. a large value of privacy level ϵ (smaller social distances equivalently) results in a lower level of indistinguishability and vice versa. However, for the sake of data utility, unnecessary randomization must be avoided such that a balance between data utility and the level of privacy protection must be kept regarding the trade-off between data utility and privacy protection.

Since the proposed mechanism publishes location data, we customize the adjacency relation defined in the standard differential privacy in order to use differential privacy framework in the location domain. This is shown in Definition 4.1.

Definition 4.1 *Adjacency relation:* Locations L and L' are considered adjacent if the distance between them is less than a predefined value D, i.e.

$$||L - L'||_2 \leq D \qquad (4.25)$$

Using Definition 4.1 we customize the standard definition of differential privacy to our needs. For this reason, we present the concept of (D, ϵ)-location privacy in Definition 4.2.

Definition 4.2 (D, ϵ)-*location privacy:* Suppose $L \in R^2$ be a user's private location and $L' \in R^2$ is adjacent to L, (i.e. $||L - L'||_2 \leq D$). Mechanism $\mathcal{A} : R^2 \to R^2$ is (D, ϵ)-location private if for any $S \subseteq Range(\mathcal{A})$ we have

$$ln\left(\frac{Pr[\mathcal{A}(L) \in S]}{Pr[\mathcal{A}(L') \in S]}\right) < \epsilon \qquad (4.26)$$

Intuitively, if an adversary wants to infer L, the distinguishability between L and any adjacent location L' that he selects is limited by ϵ. In other words, all adjacent locations L' have an equal chance to be placed in the region where $\mathcal{A}(L)$ is located. Therefore, the level of distinguishability is determined by the privacy level ϵ. To simplify the notions , in the rest of this section, we simply use notion "ϵ-differential privacy" instead of "(D, ϵ)-location privacy".

Now suppose $L_i \in R^2$ is the GPS coordinates of user u_i's real location, i.e. $L_i =< L_i^{(1)}, L_i^{(2)} >$. If $d_{ij} \in R_+$ is the social distance between user u_i and u_j, then using mechanism $M : R^2 \to R^2$ to generate response L_{ij} that user u_j receives as an approximation of user u_i's location.

$$L_{ij} = M(L_i, \epsilon(d_{ij})) = L_i + N(\epsilon(d_{ij})), \qquad (4.27)$$

where $\epsilon(d_{ij})$ is the privacy level required by user u_i against user u_j and N is a two-dimensional Laplace-distributed random variable with scale $\epsilon(d_{ij})$.

It can be concluded that the accuracy of the response L_{ij} depends on the amount of injected noise $N(\epsilon(d_{ij}))$. The noise level itself is determined by the privacy level ϵ (which is the scale of N's distribution) because the probability density function of the Laplace distribution states that a smaller amount of noise is generated with high probability for larger values of ϵ and vice versa. Therefore, since ϵ is an inverse function of d_{ij}, we can say that the system generates a more accurate response for friends with smaller social distance (or larger privacy level equivalently) while casual friends and strangers receives more generalized responses.

We already mentioned three properties for the system, i.e. *flexible (variable) privacy*, *independent of user collaboration*, and *optimal accuracy*. Regarding the first property, we can say that the system offers variable privacy because users with different social distances from a specific user receive responses with different accuracy. This accuracy has an inverse relation with the social distance between the users.

Therefore, the system provides a variable privacy protection tool for social network users to preserve their location privacy against a spectrum of users (from family members and close friend to strangers). Moreover, it is independent of user collaboration. The reason is that, the system always considers a default privacy protection plan for all users by taking function f' with a default value for constants c_1 to c_3. Therefore, there is always a default privacy plan for each user even if she is not aware of such a privacy protection tool.

After the distance to privacy function is determined, the noise injection mechanism is executed independent of user collaboration. This is applied even to non-competent users who can not collaborate with privacy protection systems due to different reasons (e.g. lack of sufficient language or computer skills) or are not aware of privacy violation risks in the social network until their privacy is violated.

Regarding the third property (i.e. *optimal accuracy*), we analyze the accuracy of system responses in terms of squared errors in the next section.

4.2.4 System Analysis

This section analyzes the performance of the system from accuracy and security perspectives. First, we assess accuracy of the responses generated by the system to ensure that it offers optimal utility. Next, the system immunity against privacy attacks is assessed. Our analysis shows that the system offers optimal accuracy which depends on ϵ only. In addition, from a security point of view, the results of our analysis indicate that the proposed system is resilient to post processing and collusion attacks.

4.2.4.1 Accuracy

It is vital for a privacy protection system to keep a balance between data utility and the level of privacy protection. To maintain data utility, the system must preserve the accuracy of privacy-aware responses. For this reason, the optimal amount of noise should be injected to the users' private location regarding the trade-off between privacy protection and data utility. In other words, the noise magnitude must not be more than what is required for privacy protection.

In the proposed mechanism, the accuracy of response L_{ij} can be measured by squared error Δ_{ij} as

$$\Delta_{ij} = ||L_{ij} - L_i||_2^2 \quad i, j \in V, \tag{4.28}$$

where a smaller error represents more accuracy. Then, we have

$$\begin{bmatrix} L_{ij}^{(1)} \\ L_{ij}^{(2)} \end{bmatrix} = \begin{bmatrix} L_i^{(1)} \\ L_i^{(2)} \end{bmatrix} + \begin{bmatrix} N_1(\epsilon(d_{ij})) \\ N_2(\epsilon(d_{ij})) \end{bmatrix}, \tag{4.29}$$

where $L_{ij}^{(k)} \in R$ $(k = 1, 2)$ are the GPS coordinates of response L_{ij} and $N_k(\epsilon(d_{ij}))$ $(k = 1, 2)$ are independent and identically distributed random variables, i.e.

$$N_k \sim Lap(0, \epsilon(d_{ij})) \qquad \forall i, j \in V, \ k = 1, 2 \tag{4.30}$$

Therefore, the squared error Δ_{ij} is obtained as

$$\Delta_{ij} = N_1^2(\epsilon(d_{ij})) + N_2^2(\epsilon(d_{ij})) \ i, j \in V \tag{4.31}$$

$N_1^2 + N_2^2 = \Delta$ corresponds to a circle with radius $\sqrt{\Delta}$, for cumulative distribution function of Δ. Therefore, we have

$$F_\Delta(\delta) = Pr[\Delta \leq \delta] = Pr[(N_1^2 + N_2^2) \leq \delta]$$
$$= \int_{-\sqrt{\delta}}^{\sqrt{\delta}} \int_{-\sqrt{\delta-n_2^2}}^{\sqrt{\delta-n_2^2}} f_{N_1, N_2}(n_1, n_2) dn_1 dn_2.$$

Since N_1 and N_2 are independent and identically distributed we have

$$f_{N_1, N_2}(n_1, n_2) = f_{N_1}(n_1) f_{N_2}(n_2) = \frac{\epsilon^2}{4} e^{-\epsilon(|n_1| + |n_2|)}. \tag{4.32}$$

Therefore,

$$F_\Delta(\delta) = \int_{-\sqrt{\delta}}^{\sqrt{\delta}} f_{N_2}(n_2) \int_{-\sqrt{\delta-n_2^2}}^{\sqrt{\delta-n_2^2}} \frac{\epsilon}{2} e^{-\epsilon|n_1|} dn_1$$
$$= \frac{\epsilon}{2} \int_{-\sqrt{\delta}}^{\sqrt{\delta}} (1 - e^{-\epsilon\sqrt{\delta-n_2^2}}) e^{-\epsilon|n_2|} dn_2$$

By taking differentiation, we obtain the probability density function (PDF) of Δ as

$$f_\Delta(\delta) = \frac{d}{d\delta} F_\Delta(\delta) = \frac{\epsilon}{2\sqrt{\delta}} e^{-\epsilon\sqrt{\delta}} \tag{4.33}$$

From above, it is derived that Δ has generalized gamma distribution [73] with scale parameter $1/\epsilon^2$, expected value $2/\epsilon^2$ and variance $20/\epsilon^4$. This means that the random variable Δ depends only on ϵ, i.e. for larger values of ϵ (equivalently, smaller social distances), with high probability, a smaller Δ is offered and vice versa. Therefore, the accuracy of the responses L_{ij} is determined by the privacy level ϵ only and they have a direct relation, i.e. any increase in ϵ results in a more accurate response. This is exactly what the mechanism needs to satisfy: *flexible (variable) privacy* and *optimal accuracy* (Fig. 4.12).

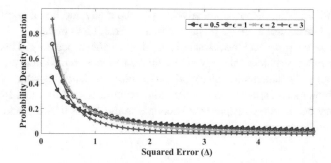

Fig. 4.12 Probability density function for generalized gamma distribution. As you see, for $\epsilon = 3$ the maximum possible error Δ is around 3, while this value is increased as ϵ decreases. This means that, with high probability, the mechanism offers less squared error Δ for larger values of ϵ

4.2.4.2 Security

In the following, we analyze the proposed system's performance against privacy attacks. For this reason, we first show that the system is immune to post processing. In other words, if an adversary has no additional knowledge about a user's real location, he cannot make the system's responses less private by performing computation on the output of the system. Next, we prove that the proposed system is resilient against collusion attacks in which a group of users collaborate and share their received responses to obtain a more accurate approximation.

Proposition 4.1 (Resilience to post processing) *If* $\mathcal{M} : R^2 \to R^2$ *is the proposed mechanism which preserves* ϵ*-differential privacy, then for any function* $f : R^2 \to R^2$*, the composition* $f \circ \mathcal{M} : R^2 \to R^2$ *also preserves* $\epsilon-$ *differential privacy.*

Proof Assume location L' is adjacent to L, i.e. $||L' - L|| \leq D$ (see Definition 4.1) and $S' \subset R^2$. By defining $S = \{l \in R^2 : f(l) \in S'\}$ and because $\mathcal{M}$ is a $\epsilon-$ differential private mechanism we have

$$Pr[f(\mathcal{M}(L)) \in S'] = Pr[\mathcal{M}(L) \in S] \leq e^{\epsilon} Pr[(L') \in S'] \qquad (4.34)$$

Therefore, according to the definition of S we obtain

$$Pr[f(\mathcal{M}(L)) \in S'] \leq e^{\epsilon} Pr[f(\mathcal{M}(L')) \in S'] \qquad (4.35)$$

which means $f \circ \mathcal{M}$ is also $\epsilon-$differential private.

Resilience to post processing is a common advantage of mechanisms that adopt the differential privacy framework. It guarantees that after the system publishes an

ϵ−differential private response, an adversary without any additional knowledge on the private data cannot increase privacy loss and make it less private.

Therefore, the proposed mechanism is resilient to post processing. This makes it immune to privacy attacks that rely solely on post processing. Moreover, we proof that the proposed mechanism is also resilient to collusion attacks in which a group of users combine their responses to make a more accurate approximation. In practice, an adversary can create multiple fake accounts in the social network and establish such a colluding group.

Theorem 4.1 (Resilience to collusion): *Consider a group of K users $C \subseteq V$ who collaborate and share their response $M(l_i, \epsilon(d_{ij})) = l_{ij} (j = 1, 2, \ldots, K), (i \in V)$ to obtain $l_c^{(i)}$. If l_{ij} be an $\epsilon(d_{ij})$-differentially private response, then $l_c^{(i)}$ is $(max_{j \in C} \epsilon(d_{ij})$-differentially private.*

Proof

$$\begin{bmatrix} l_{i1} \\ l_{i2} \\ \cdot \\ \cdot \\ \cdot \\ l_{iK} \end{bmatrix} = l_i + \begin{bmatrix} N(\epsilon(d_{i1})) \\ N(\epsilon(d_{i2})) \\ \cdot \\ \cdot \\ \cdot \\ N(\epsilon(d_{iK})) \end{bmatrix}, \tag{4.36}$$

where l_i is the private location of user u_i. We sort the responses $L_{ij} (j = 1, 2, \ldots, K)$ such that

$$\epsilon(d_{i1}) < \epsilon(d_{i2}) < \cdots < \epsilon(d_{iK}), \tag{4.37}$$

which means l_{iK} is the most accurate response among $l_{ij} (j \in C)$. To obtain $l_c^{(i)}$, the adversary combines K received responses l_{ij}. Therefore,

$$l_c^{(i)} = \sum_{j=1}^{K} w_j l_{ij} = \sum_{j=1}^{K} w_j (l_i + N(\epsilon(d_{ij}))), \tag{4.38}$$

where w_j is the weight considered for response j in the combination process. For simplicity we assume $w_j (j \in C)$ are equal, i.e.

$$w_j = \frac{1}{K} \quad j = 1, 2, \ldots, K. \tag{4.39}$$

Therefore,

$$l_c^{(i)} = l_i + \frac{1}{K} \sum_{j=1}^{K} N(\epsilon(d_{ij})) \tag{4.40}$$

By defining $N(\epsilon(d_{ij})) = N_{ij}$ we have

$$l_c^{(i)} = l_i + \frac{1}{K} \sum_{j=1}^{K} [N_{iK} + \sum_{m=j+1}^{K} (N_{im-1} - N_{im})]. \tag{4.41}$$

Since $l_i + N_{iK} = l_{iK}$ we obtain

$$l_c^{(i)} = l_{iK} + \sum_{j=1}^{K} \sum_{m=j+1}^{K} (N_{im-1} - N_{im}) \tag{4.42}$$

We can say that $l_c^{(i)}$ consists of two parts. First, ϵ_{iK}-differential private l_{iK} which is the most accurate response in C since $\epsilon_{iK} = max_{j \in C} \epsilon_{ij}$ and second, a noise section. Since N_{im-1} and N_{im} are independent Laplace-distributed random variables, $(N_{im-1} - N_{im})$ has also Laplace distribution. Therefore, we can consider $\phi = \sum_{j=1}^{K} \sum_{m=j+1}^{K} (N_{im-1} - N_{im})$ as Laplace-distributed noise added to l_{iK}. In conclusion, we can say that $l_c^{(i)}$ is the ϵ_{iK}-differential private response l_{iK} which has been post processed by function $g(x) = x = \phi$, i.e.

$$l_c^{(i)} = g(l_{iK}) = g(M(l_i)) = l_{iK} + \phi. \tag{4.43}$$

According to Proposition 4.1, mechanism M is immune to post processing, hence, $l_c^{(i)}$ is also ϵ_{iK}−differential private. Therefore, the result of any collusion attack is equivalent to a $(max_{j \in C} \epsilon(d_{ij}))$-differential private response which means no more accuracy is obtained. consequently, there is no need for additional privacy preserving noise when multiple users ask for a user's private location.

4.2.5 Performance Evaluation

In this section, we evaluate the performance of our proposed DBLP2 system. Firstly, we evaluate the proposed system's performance regarding the four types of users, i.e. conservative user, very relaxed user, relaxed user, and moderate user. Finally, we assess the immunity of the proposed system against collusion attacks in Sect. 4.1.3.

4.2.5.1 Variable Privacy

To evaluate the system performance in terms of variable privacy, a single user scenario is considered in which the user's location privacy is protected against a variety of users. For this reason, we assess the magnitude of the injected noise for a spectrum of users (i.e. from family members and close friends to casual friends and strangers). To model this scenario, we increase the social distance d from 0 to ∞ and obtain

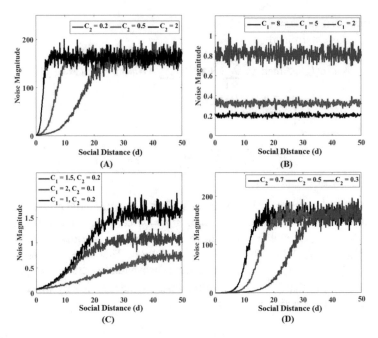

Fig. 4.13 The magnitude of the injected noise for a **a** conservative user, **b** very relaxed user, **c** relaxed user, and **d** moderate user

the related privacy level ϵ using the function. Then, for each value of the obtained privacy level ϵ, the magnitude of the injected Laplace noise is calculated.

By selecting the appropriate values for constants c_1, c_2 and c_3, the distance to privacy function f' can model the behavior of different users in choosing a privacy protection policy. Therefore, we adopt this function for the experiments to convert social distances to privacy levels. In this regard, the behavior of the four types of users introduced in Sect. 4.3 are modeled using this function by selecting the suitable values for c_1, c_2 and c_3. Finally, based on the privacy levels obtained, the related noise magnitude is calculated. The results of our experiments are shown in Fig. 4.13.

For the first type of user, i.e. the conservative user, the result shows that the amount of injected noise is largely increased when the social distance is raised above zero. This means that the system generates responses with high accuracy for the user's family members and close friends (who have small social distance) while other users receive a totally inaccurate response. We performed the experiments for three different values of constant c_2 to see the effect of this parameter. As you see, c_2 determines the threshold social distance at which a tight privacy protection (required by the user) starts. In other words, c_2 represents how a user is conservative. We have also selected $c_1 = 0$ and $c_3 = 0.1$ in this case (c_1 must be zero for this type of user).

Figure 4.13b shows the result for a very relaxed user ($c_2 = c_3 = 0$). In this case, c_1 determines a high privacy level (relaxed privacy) which the user selects against all the other users. As you see, the system always generates a very small noise regardless of the social distance (a relatively accurate response for all the other users). The level of this noise is determined by c_1. In other words, for larger c_1 (higher privacy level) a more accurate response is generated. You can realize the difference between the system responses generated for the first and second type of users, if you compare the amount of the noise generated for each category. Moreover, the amplitude of the changes in the amount of generated noise is higher for a smaller c_1. The reason is that the variance of the generated Laplace noise is $2/\epsilon^2$. Thus, the variance is increased as c_1 is decreased.

The noise magnitude for the third type of user, i.e. the relaxed user, is shown in Fig. 4.13c for different values of c_1 and c_2 ($c_3 = 3$ is selected in this case). The noise magnitude for this type of user is almost the same as what we have in Fig. 4.13b (notice the amount of noise magnitude in Fig 4.13b and c). The only difference is that in this case, the user requires less privacy protection for small social distances while in the second type, there is no difference between different social distances in terms of privacy protection.

Finally, for the last type of users, i.e. the moderate user, which we propose her behavior as the standard behavior, the result is shown in Fig. 4.13d for different values of c_2 ($c_1 = 0, c_3 = 3$). As you see, for small social distances, the system generates accurate responses (the noise magnitude is very small) while the level of accuracy is gradually increased as the social distance gets bigger. The constant c_2 determines the slope of this increment such that for a bigger c_2, the noise magnitude is increased with a higher rate.

4.2.5.2 Collusion Attacks

In this part, we consider a collusion attack in which five users share their received responses to obtain a more accurate approximation of the victim's location. In practice, an adversary can establish such a colluding group by creating five fake accounts in the social network. We assume that these five users have different social distances from the victim. In other words, the victim has different privacy levels $\epsilon_1, \epsilon_2, \ldots, \epsilon_5$ against these five users. Hence, they receive responses with different accuracy as well, i.e. the user with the largest ϵ (smallest social distance) receives the most accurate response and vice versa. We compare the accuracies of these responses and the collusion outcome to see if there exists any motivation for an adversary to perform a collusion attack or not. In order to have a better picture of the system performance, we have performed the experiments 50 times for each user and obtained the squared error Δ of the five responses and the outcome of the collusion in each iteration. You can see the result in Fig. 4.14.

As we discussed in last subsection, the resultant squared error is a random variable with the generalized gamma distribution. Therefore, as Fig. 4.14 shows, a different error has been obtained for a specific user for separate experiments. In addition, the amplitude of these changes is higher in a response with a lower ϵ. The reason is that the variance of the squared error Δ, i.e. $20/\epsilon^4$, is larger for a lower ϵ. Moreover, as we expect, the accuracy of each response only depends on the privacy level $\epsilon(d)$.

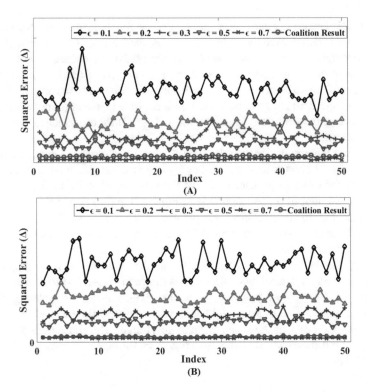

Fig. 4.14 The result of a collusion attack in which five users with different social distances from the victim have shared their response to obtain a more accurate location data

Consequently, in each iteration, the user with the lowest ϵ has received the response with the largest error and vice versa.

You see in Fig. 4.14 that the outcome of the collusion attack is almost the same as the most accurate response and has never been more accurate than it. This confirms the results of our analysis which state that the result of any collusion attack is equivalent to a $(max_{j \in C} \ \epsilon(d_{ij}))$-differential private response. Consequently, there is no logical motivation for an adversary to conduct such a collusion attack since no additional benefit can be gained.

Moreover, regarding the combination process of the responses, Fig. 4.14a shows the result of the experiments when the same weights have been considered for the responses, i.e. the responses have equal share in creating the collusion result. However, the results shown in Fig. 4.14b is for a case in which the responses have been combined with different weights. As you see, the same result has been obtained in both cases which confirms that the combination process does not affect the immunity of the system against collusion attacks.

4.3 Personalized Privacy in Social Networks Using Anonymity Based Methods

In this section of the book, we investigate anonymity as another approach to provide privacy in social networks. As far as we know, social networks do not offer publicly available anonymous group messaging. If these services are employed by social network service providers, users can be able to create a group in social media and anonymously post their opinions. For example, consider a group of journalists; each journalist wishes to publish some secret government information that he/she obtained from a confidential source. Using an anonymous communication protocol, they can create a group in social media and anonymously publish their posts without the risk of prosecution. Note that the protocol not only needs to hide the origin of each post, but it must also be resistant against traffic analysis to prevent a government agency or an ISP from identifying the message publishers by monitoring and analysing the journalists' traffic in the network.

To offer anonymity, several anonymous communication networks (ACNs) have been proposed so far such as onion routing [74], AN.ON [75], and Tor [76]. These solutions work based on mixnet [77], a basic anonymous communication protocol. However, to guarantee users' anonymity, they need access to a set of geographically distributed servers (or mixes) such that at least some of them are trusted [78]. In addition, mixnet-based networks cannot provide the necessary protection against traffic analysis attacks [79–82]. These attacks can be conducted by powerful adversaries like large ISPs who can monitor users' traffic in the network [80]. Dining Cryptographers network (DC-net) [83] is another anonymous communication protocol that guarantees protection against traffic analysis attacks. Unlike mixnet, DC-net is completely performed by the users themselves and does not require any proxy. However, DC-net suffers from three critical issues that reduce its practicality. Firstly, there is a collision possibility issue. Users' messages are exposed to corruption due to possible collisions. In DC-net, every user publishes a vector of data that has N elements (positions or slots) where N is the number of users in the group who want to anonymously publish a message. It requires every user to place his/her message in a unique slot where other users must insert their keys XORed together. Any deviation from this procedure makes all the users' messages unrecoverable. Secondly, DC-net is vulnerable to disruptions and Denial of Service (DoS) attacks since a malicious user can disrupt the protocol by sending irrelevant bit-streams in each of the N slots.

Finally, DC-net is able to provide anonymity only for a few protocol cycles. We name this issue the Short stability problem. To the best of our knowledge, no previous research work has identified this flaw in the DC-net performance. We prove that it is feasible to infer the origin of each message, after users published their messages for at least three protocol cycles.

4.3.1 Literature Review

In this section, we present a brief review of the literature on anonymous communication protocols. Prior significant research work in this area started in the early 1980s when Chaum presented mixnet [77]. In mixnet, users' encrypted messages are batched together and successively relayed into the network after they are decrypted and shuffled by a set of proxies (named as mixes). Several extensions of the mixnet protocol have been proposed so far such as onion routing [74], Tor [76], and An.On [75]. However, these protocols require that users' messages are passed through a series of proxies which results in high latency and makes them vulnerable to traffic analysis. Moreover, they are vulnerable to active attacks and disruptions which break the anonymity guarantees and cause protocol jamming, respectively [80]. In addition, mixnet-based protocols offer anonymity as long as at least one mix in the network executes the protocol honestly.

Beside the original mixnet, Chaum introduced DC-net as another option towards anonymous communication [83]. DC-net is a distributed and non-interactive protocol that allows a group of users to anonymously publish their messages in a single broadcast round. It provides users with secure anonymous communication if the protocol is executed honestly. However, DC-net suffers from three critical issues that make it impractical [78, 84].

To address the DC-net problems, a number of DC-net extensions have been introduced. Dissent [84] focuses on addressing traffic analysis and DoS attacks to which mixnet and DC-net protocols are vulnerable. For this reason, the authors of Dissent have proposed a mechanism to trace disrupting (misbehaving) users. This is called accountability in the literature. However, in Dissent, the employed shuffling mechanism imposes a delay at the start of each round that makes the protocol impractical for delay-sensitive applications.

Herbivore [85] is another anonymous group messaging protocol that provides anonymity by dividing a large group of network users into smaller DC-net subgroups. In fact, in Herbivore, the size of user groups is reduced in order to limit the attack surface. This enables the protocol to provide only small sizes of anonymity sets.

Although DC-net-based protocols have a decentralized and non-interactive structure, a few numbers of server-based protocols have also been proposed in the literature. For example, Wolinsky et al. [86] suggest a client/server architecture to achieve a high level of scalability. In their proposed protocol many untrusted clients anonymously publish their messages through a smaller and more trusted set of servers. The protocol offers traffic analysis resistance and strong anonymity, provided that there is at least one honest server. However, the proposed disruptor tracing procedure is too costly. To solve this issue, public-key cryptography and zero-knowledge proofs are used in Verdict [80] to infer and exclude any misbehaviour before it results in a disruption. However, no security analysis has been presented in the paper to proof its security. Riffle [81] is another server-based protocol proposed in this area of research. It consists of a small set of servers that provide anonymity for a group of users. However, it still relies on at least one trusted proxy server.

Prior work on privacy issue of Location-Based Services has mostly focused on *K-Anonymity* and *Dummy-Based* methods although some efforts have recently done on other techniques such as *Differential Privacy*, and *Cryptography-Based* schemes.

K-Anonymity efforts [87–90] require a trusted third-party server which is called an anonymizer, between users and LSP. The anonymizer receives service requests from a user and enlarges its location into a region (cloaking region) so that it contains the locations of K-1 other users as well as location of the requesting user. Therefore, the adversary cannot identify the requesting user among other K-1 users. The advantage of these methods is that the communication cost between users and anonymizer is reduced, however, they suffer from decreased QoS because when there are not enough users near the requested user, the anonymizer has to increase the radius of cloaking region, hence, the increased processing times results in a greater service latency. To solve this problem, some efforts have been done in [89, 90] to increase QoS. In these papers the area of cloaking region is minimized by using footprints-historical locations of other users.

Several dummy-based location privacy schemes have been proposed so far for location privacy protection. In all of them users send their location data including noise (some fake location data or dummies) to LSP directly. Thus, there is no need to a trusted anonymizer. In [91], a dummy generation algorithms have been presented, which is *Moving in a Limited Neighbourhood*. In this work, the dummies are generated in a neighbourhood of the previous position of the dummies. Also, a cost reduction technique was proposed in [91] to limit the communications overhead caused by sending dummies. However, generating dummies at random or through a fixed rule can not provide flexible location privacy for users. Hence, in [92], a Privacy-Area Aware scheme is proposed based on a flexible dummy generation algorithm in which dummies are generated according to either a virtual grid or circle. This approach provides configurable and controllable dummy generation by which it is possible to control the user's location privacy. But a disadvantage of this method is that it doesn't consider nature of the region. For example, some dummies may be generated in places which are unlikely for a user to be there (e.g., in a river). To solve this problem, in [93] a Dummy-Location Selection (DLS) method has been proposed to prevent the adversary from exploiting side information such as a region map. This is done by carefully selecting dummies based on the entropy metric.

However, in [64, 94] it has been showed that when a user adopts one of the aforementioned dummy-based methods, the adversary can identify some dummies with a minimum correct ratio of 58% by means of the spatiotemporal correlation between neighbouring location sets. Therefore, they have proposed a Spatiotemporal Correlation-Aware privacy protection scheme in which correlated dummies are filtered out and only uncorrelated dummies are sent to LSP. But this method can protect user's location privacy under some conditions only and if the adversary estimates the threshold angle which is used to filter space correlated dummies, he will be able to identify dummies or even the user's real location.

4.3.2   Preliminaries

This section presents the foundation for the next sections. After briefly reviewing DC-net protocol, we introduce its drawbacks and explain why it requires modifications.

4.3.2.1   DC-Net Overview

DC-net [83] is a distributed and non-interactive protocol proposed to provide anonymous communications for a group of users who wish to anonymously publish their messages in the group. Its title comes from the example by which Chaum explained his proposed protocol (Fig. 4.15).

Three cryptographers sit around a table in a restaurant to have dinner. They are informed by a waiter that someone has anonymously paid their bill. The payer can be one of them or the bill might have been paid by the NSA (National Security Agency). They respect each other's right to make an anonymous payment, but they are curious to see if NSA has paid the bill. Thus, they perform the following protocol.

For all pairs, two cryptographers share a secret bit by tossing an unbiased coin behind their menu such that only those two cryptographers see the result. Thus, cryptographer A, for example, has two secret bits k_{AB} and k_{AC} that have been shared with cryptographer B and C, respectively. Then, if a cryptographer has paid the bill, he XORs his shared keys with bit 1. Otherwise, the XOR operation is performed with bit 0. In both cases, each cryptographer announces his result. If the three published

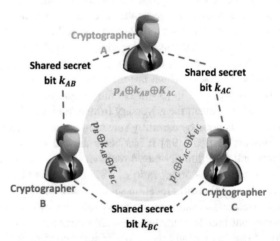

Fig. 4.15 The dining cryptographers network in a simple example

$$p = (p_A \oplus k_{AB} \oplus K_{AC}) \oplus (p_B \oplus k_{AB} \oplus K_{BC}) \oplus (p_C \oplus k_{BC} \oplus K_{AC})$$

$$= p_A \oplus p_B \oplus p_C = \begin{cases} 0, & \text{NSA has paid the bill} \\ 1, & \text{Otherwise} \end{cases}$$

results are XORed together, the result bit is 0 if NSA has paid their bill. If one of the cryptographers has paid the bill, the result is 1.

This basic protocol has been extended to multiple users in [95]. Let's consider N users $u_1, u_2, u_3, \ldots, u_N$ who wish to anonymously publish some L-bit messages $\mathbf{m}_i$ ($i = 1, 2, \ldots, N$). Assume that each pair of users (u_i, u_j) shares an L-bit key $\mathbf{k}_{ij}(w)$ in a set-up phase where $\mathbf{k}_{ij}(w) = \mathbf{k}_{ji}(w)$ for $i, j, w \in \{1, 2, \ldots, N\}$. Moreover, in this phase, every user computes the following *XORed Keys* (XK) vector:

$$\mathbf{X}_i = [\mathbf{x}_i(1) \ \ \mathbf{x}_i(2) \ \ \mathbf{x}_i(3) \ \ldots \ \mathbf{x}_i(N)], \tag{4.44}$$

where

$$\mathbf{x}_i(w) = \oplus_{\substack{j=1 \\ j \neq i}}^{N} \mathbf{k}_i j(w), \quad w = 1, 2, \ldots, N. \tag{4.45}$$

After the set-up phase, users can broadcast their messages by performing the following steps:

(1) Every user u_i randomly selects a slot (position) $s_i \in \{1, 2, \ldots, N\}$ in his/her $\mathbf{X}_i$ vector.

(2) The XK vector $\mathbf{X}_i$ is converted to $\mathbf{Y}_i$ by replacing $\mathbf{x}_i(s_i)$ with $\mathbf{m}_i \oplus \mathbf{x}_i(s_i)$. Then, $\mathbf{Y}_i$ is published.

Since $\oplus_{i=1}^{N} \mathbf{x}_i(w) = 0$ for $w = 1, 2, \ldots, N$, if users have selected different positions, we have:

$$\oplus_{i=1}^{N} \mathbf{Y}_i = \mathbf{M}', \tag{4.46}$$

where $\mathbf{M}'$ is the users' messages vector $\mathbf{M} = [\mathbf{m}_1 \ \ \mathbf{m}_2 \ \ \mathbf{m}_3 \ldots \mathbf{m}_N]$ in which the elements have been shuffled. We define $\mathbf{M}'$ as the *Shuffled Messages Vector* (SMV) since we need to refer to this vector frequently.

Therefore, the users' messages are published for the group in such a way that the origin of each message is anonymous.

4.3.3 DC-Net Drawbacks

Although DC-net offers strong anonymity, it suffers from some critical issues:

- *Collision possibility*: In DC-net, it is assumed that the users select different slots (or positions) in the XK vector $\mathbf{X}_i$. However, if two users u_i and u_j place their messages in the same slot (i.e., $s_i = s_j$ are selected by them), $\mathbf{m}_i \oplus \mathbf{m}_j$ is recovered in the $\mathbf{M}'$ vector at the final stage that makes both $\mathbf{m}_i$ and $\mathbf{m}_j$ unrecoverable (note that in this case, one element of $\mathbf{M}'$ is obtained as $\mathbf{0}$).
- *Vulnerability against disruptions* (security issue): DC-net works well only when users execute the protocol honestly. The protocol is jammed if a malicious user, for example, fills $\mathbf{Y}_i$ with some random bits and publishes it. In this case none of the users' messages is successfully recovered.

Apart from that, we identified another critical issue, i.e. short stability, in the DC-net performance which is discussed in the next subsection.

4.3.4 The Short Stability Issue

We noticed that DC-net provides anonymity only for a few protocol cycles. After users publish their messages for at least three cycles, it is possible to infer the origin of each message by analysing vectors Y_i published in the previous three cycles by the users. To clarify this, consider the following example:

DC-net is performed by a group of four users u_1, u_2, u_3, and u_4 who want to publish some 5-bit messages. They publish $Y_i^{(1)}, Y_i^{(2)}$, and $Y_i^{(3)}$ $(i = 1, 2, 3, 4)$ in the first three protocol cycles. Suppose the XK vector for user u_1 is $X_1 =$ [11000 10100 00110 10110], and for these three cycles, he/she selects slots 2, 4, and 1 in the XK vector X_1 to XOR his/her messages $m_1^{(1)} = 10011, m_1^{(2)} = 11001$, and $m_1^{(3)} = 10101$, with X_1 components, respectively. Thus, for the published vector Y_1 we have:

$$Y_1^{(1)} = [11000\ 00111\ 00110\ 10110]$$
$$Y_1^{(2)} = [11000\ 10100\ 00110\ 01111]$$
$$Y_1^{(3)} = [01101\ 10100\ 00110\ 10110]$$

By analysing these three vectors, the XK vector X_1 is easily obtained. Intuitively, if different slots are selected by the user u_1, for a specific $w \in \{1, 2, 3, 4\}$, $x_1(w)$ is the element in the set $\{y_1^{(l)}(w)\}$ $(l = 1, 2, 3)$ that has been repeated at least twice. Having X_1, the other users are able to compute $X_1 \oplus Y_1^{(j)}$ and identify u_1 as the publisher of $m_1^{(1)}, m_1^{(2)}$, and $m_1^{(3)}$. If the same slot is chosen for at least two cycles, the elements of the above set are totally different (assuming there are different messages in each cycle). In this case, this slot is identified as the one in which the user has XORed his/her message during at least two of these cycles.

4.3.5 HSDC-Net: Secure Anonymous Messaging in Online Social Networks

In this section, we propose Harmonized and Stable DC-net (HSDC-net), a self-organizing protocol for anonymous communications. In our protocol design, we first resolve the short stability issue and obtain SDC-net, a stable extension of DC-net. Then, we integrate the Slot Reservation and Disruption Management sub-protocols into SDC-net to overcome the collision and security issues, respectively. Our prototype implementation shows that HSDC-net achieves low latencies that makes it a practical protocol (Fig. 4.16).

Suppose a group of N users who want to anonymously publish their messages in the group. Assume they all use a simple messaging application that does not

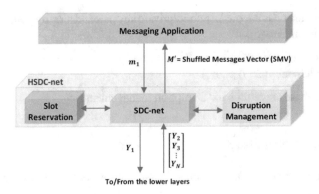

Fig. 4.16 HSDC-net system architecture

offer anonymity. We add HSDC-net (as a separate and independent module) to the messaging application to make it an anonymous message exchanging application. In this scenario, HSDC-net delivers the SMV to the messaging application in which the other users' messages are placed in such a way that the origin of each message is anonymous.

The proposed protocol is performed in three phases (1) *Initialization*, (2) *Scheduling*, and (3) *Message Publishing* (MP). In the scheduling phase, after the protocol initialization, the available N slots are anonymously allocated to the users. Then, they publish their messages by continuously executing the MP phase. In the following, we present each phase individually and in the order that they are performed.

4.3.5.1 Initialization

In this phase, all the N users in the group execute an initialization algorithm individually. Considering user u_1, this algorithm takes as input the user vector $U = [u_2 \ u_3 \ \ldots u_N]$ and generates the following items in collaboration with the users specified in U:

- A matrix of pairwise symmetric keys $\mathbf{K}_1^{(0)} = \begin{bmatrix} \mathbf{K}_{12}^{(0)} \\ \mathbf{K}_{13}^{(0)} \\ \ldots \\ \mathbf{K}_{1N}^{(0)} \end{bmatrix}$,

  in which $\mathbf{K}_{1j}^{(0)} = [\mathbf{k}_{1j}^{(0)}(1) \ \mathbf{k}_{1j}^{(0)}(2) \ \ldots \ \mathbf{k}_{1j}^{(0)}(N)]$, where $\mathbf{k}_{1j}^{(0)}(w), (w = 1, 2, \ldots, N)$ is an L-bit secret symmetric key that u_1 shares with u_j $(j = 2, 3, \ldots, N)$ to use in slot w.
- Vector $R_1 = [r_{12} \ r_{13} \ \ldots r_{1N}]$, where r_{1j} is a random integer number shared secretly between users u_1 and u_j.

Moreover, by executing this algorithm, each user u_i signs the following two items using his/her private key and sends them to user u_j $(j = 1, 2, \ldots, N, j \neq i)$ (we assume every user has adopted a public/private key pair and already published his/her public key in the network):

- The jth row of matrix $\mathbf{K}_i^{(0)}$ (i.e., $\mathbf{K}_{1j}^{(0)}$ for user u_1).
- The jth element in R_i (i.e., r_{ij}).

We will use these signatures for disruption management.

4.3.5.2 Scheduling Phase

In this phase, every user u_i performs the SR sub-protocol that is executed in the following steps:

(1) u_i creates vector $\mathbf{S}_i = [\mathbf{S}_i(1) \ \mathbf{S}_i(2) \ldots \mathbf{S}_i(N)]$, in which every element consists of L zero bits (i.e., $\mathbf{S}_i(w) = \mathbf{0}$ for $w = 1, 2, \ldots, N$).

(2) Two random integer numbers l and n are selected by u_i in $[1, L]$ and $[1, N]$, respectively. Then, the lth bit in $\mathbf{S}_i(n)$ is set to 1 to obtain $\mathbf{S}_i'$ (assuming $\mathbf{S}_i'$ is a single bit-stream, it has only a single bit 1 in the position $l + (n-1)L$).

(3) u_i computes and publishes the following vector $\mathbf{Z}_i$:
$$\mathbf{Z}_i = [\mathbf{Z}_i(1) \ \mathbf{Z}_i(2) \ldots \mathbf{Z}_i(N)],$$
where $\mathbf{Z}_i(w) = [\oplus_{\substack{j=1 \\ j \neq i}}^{N} \mathbf{k}_{ij}^{(0)}(w)] \oplus \mathbf{S}_i'(w)$, $w = 1, 2, \ldots, N$.

(4) Upon receiving $N - 1$ vector $\mathbf{Z}_j$ ($j = 1, 2, \ldots, N, j \neq i$) from the other users (who performed the same procedure), u_i computes vector $\mathbf{V} = [\mathbf{V}(1) \ \mathbf{V}(2) \ldots \mathbf{V}(N)]$, in which $\mathbf{V}(w) = \oplus_{i=1}^{N} \mathbf{Z}_i(w)$. Note that $\mathbf{V}(w) = \oplus_{i=1}^{N} \mathbf{S}_i'(w)$ because the terms related to the pairwise keys are cancelled out when they are XORed together. Thus, if we consider vector $\mathbf{V}$ as a single bit stream, it shows all the 1 bits set by the users (in step 2) placed in their primary positions (Fig. 4.17).

(5) u_i computes the hamming weight of $\mathbf{V}$, i.e., $H = Hamming(\mathbf{V})$ that indicates how many bits 1 exist in $\mathbf{V}$. Based on the obtained H, two situations are supposable:

Fig. 4.17 A simple illustration of SR performance

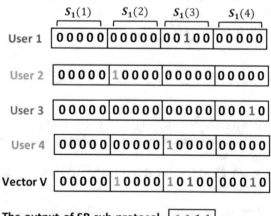

- If $H = N$, there is no collision and every user has selected a unique slot. In this case, (considering $\mathbf{V}$ as a single bit stream) u_i highlights his/her selected 1 (set in step 2) in $\mathbf{V}$, keeps all the 1s and removes all the 0s of $\mathbf{V}$. This results in a bit stream of size N in which all the N bits are 1. In this bit stream, the bit number associated with the position in which the highlighted bit 1 has been placed is the slot assigned to u_i.
- If $H \neq N$, it means two or more users have selected the same random numbers l and n in step 2. In this case, the SR sub-protocol needs to be restarted. However, to protect users against the DC-net short stability issue, we need to change the users' pairwise keys. To do this, every user u_i changes his/her symmetric keys $\mathbf{K}_i^{(0)}$ to $\mathbf{K}_i^{(1)}$ by adding r_{ij} to all the elements in the jth row of $\mathbf{K}_i^{(0)}$, $(j = 1, 2, \ldots, N, j \neq i)$. Using this technique, users can non-interactively obtain a new set of pairwise keys without imposing any further communication overhead.

As you see, we consider $\mathbf{S}_i'$ as a single LN-bit vector in which every bit represents a slot. By doing this, the N users have LN slots to select from instead of only N slots. This can reduce the probability of collision to a negligible level if L is a relatively large number.

4.3.5.3 Message Publishing (MP) Phase

After the N available slots of SMV are allocated to the N users in the scheduling phase, every user u_i can anonymously publish his/her messages. This can be done by performing the original DC-net protocol. However, as discussed before, the short stability issue must be addressed first. For this reason, we propose Stable DC-net (SDC-net) that addresses this issue.

Stable DC-net (SDC-net): In SDC-net, users change their pairwise keys before they start a new round of the MP phase (publishing a new message is done with a new set of pairwise keys). This makes the elements of $\mathbf{Y}_i^{(p)}$ dissimilar for different rounds ($p = 0, 1, 2, \ldots$).

Suppose u_i wants to publish his/her message $\mathbf{m}_i^{(p)}$ in round p of the MP phase. To change his/her pairwise keys, u_i simply adds r_{ij} (which has been secretly shared with user u_j in the initialization phase) to all the elements in the jth row of $\mathbf{K}_i^{(p-1)}$ ($j = 1, 2, \ldots, N, j \neq i$). This results in $\mathbf{K}_i^{(p)}$ which is a set of different keys in comparison to $\mathbf{K}_i^{(p-1)}$. Therefore, by applying a different set of keys, a different XK vector $\mathbf{X}_i$ is obtained in each round of MP that makes $\mathbf{Y}_i^{(p)}$ completely dissimilar to $\mathbf{Y}_i^{(p-1)}$. Note that, the new sets of pairwise keys are obtained by the users without imposing any further communication overhead to the protocol.

Now, let's return back to explain the MP phase. In round p of this phase, user u_i publishes message $\mathbf{m}_i^{(p)}$ by invoking algorithm $SDC - net(U, \mathbf{K}_i^{(p)}, \mathbf{m}_i^{(p)}, slt_i)$. This algorithm takes as input, the vector U of $N - 1$ users, matrix of pairwise keys $\mathbf{K}_i^{(p)}$, user's message $\mathbf{m}_i^{(p)}$, and the slot number assigned to u_i. The output of this algorithm is the $\mathbf{Y}_i^{(p)}$ vector that its elements are obtained using

$$\mathbf{y}_i^{(p)}(w) = \oplus_{\substack{j=1 \\ j \neq i}}^{N} \mathbf{k}_{ij}(w), \quad for \quad w = 1, 2, \ldots, N, w \neq slt_i, \tag{4.47}$$

and

$$\mathbf{y}_i^{(p)}(slt_i) = \mathbf{m}_i^{(p)} \oplus (\oplus_{\substack{j=1 \\ j \neq i}}^{N} \mathbf{k}_{ij}(slt_i)) \tag{4.48}$$

Similar to DC-net, every user is able to obtain the SMV by XORing the received $N - 1$ vector $\mathbf{Y}_j^{(p)}$ ($j = 1, 2, \ldots, N, j \neq i$) with his/her own vector $\mathbf{Y}_i^{(p)}$, i.e.:

$$SMV^{(p)} = \oplus_{i=1}^{N} \mathbf{Y}_i^{(p)} \tag{4.49}$$

As we mentioned before, users' messages are shuffled in SMV such that the origin of each message is unknown.

4.3.5.4   Disruption Management

As we discussed before, the original DC-net protocol is jammed if one or more users perform dishonestly. Since misbehaviours of dishonest users are inherently unavoidable, protecting DC-net against disruptions and jamming attacks is difficult and imposes additional time and communication overheads on the protocol [80]. Therefore, creating accountability is a good solution to address this issue.

After a disruption is detected (if the users' messages in SMV have been corrupted), assuming the disruption is detected in round p, the following steps are performed by every user u_i who detects the disruption:

(1) u_i publicly informs other users that his/her message in slot slt has been corrupted (note that revealing slt does not jeopardize u_i's anonymity since other users cannot see his/her real message which has been corrupted).

(2) Upon receiving u_i's announcement, other users publish the set of their keys related to this slot, i.e., u_j ($j = 1, 2, \ldots, N, j \neq i$) publishes $\{\mathbf{k}_{jl}^{(p)}(slt)\}_{\substack{l=1 \\ l \neq j}}^{N}$.

(3) Every user u_j checks the other users' published keys to see if a user (say u_l) has published a key different than their shared pairwise key $\mathbf{k}_{jl}^{(p)}$.

(4) If u_j (in step 3) finds that user u_l has published a key different than their shared pairwise key (i.e. $\mathbf{k}_{jl}^{(p)}$), he/she announces u_l's identity as the disruptor. To support his/her claim, u_j publishes u_l's signature on the real $\mathbf{k}_{jl}^{(p)}$ and r_{jl} received during the initialization phase.

(5) u_i computes $D_j = \oplus_{\substack{l=1 \\ l \neq j}}^{N} \mathbf{k}_{jl}^{(p)}(slt)$ for $j \neq i$.

(6) If $D_j \neq \mathbf{Y}_j^{(p)}(slt)$, u_j's identity is published by u_i as the disruptor. Other users can also confirm this by computing D_j and comparing it with $\mathbf{Y}_i^{(p)}(slt)$.

(7) The messaging application is notified by sending the identity of the disruptor(s).

After the identity of disruptor(s) is publicly announced, the users can resume the protocol, this time by excluding the disruptor(s). To do this, they need to update their matrix of pairwise symmetric keys $\mathbf{K}_i^{(p)}$ by eliminating the row(s) associated with the disruptor(s). However, they do not need to perform the initialization phase and set up new pairwise keys as the previous keys are still valid.

4.3.5.5 Multiple Reservations

In HSDC-net, it is possible that a user reserves more than one slot in the scheduling phase. The reason is that during the scheduling phase, the users have LN slots to select from which is much larger than the number of users in the group, i.e. N, specifically, if L is a large number. This is an advantage for users with a high activity rate that need to publish more messages during a single cycle. To reserve B slots ($B > 1$) when performing the SR sub-protocol, a user needs to repeat step 2 of SR for B times. Note that in this case, the users' XK vectors and SMV have $N + A(B - 1)$ elements (or slots), where A is the number of users who reserve B slots. On the other hand, it is required to consider an upper limit on the number of slots that every user can reserve. This protects the protocol from performance degradation caused by collisions.

4.3.6 Security Analysis

In this section, we show how the proposed protocol performs against different security threats. The target of these security threats can be either deanonymizing users' messages or disrupting the protocol performance.

DoS attacks on SR sub-protocol: Suppose a malicious user u_D reserves many slots by setting the majority (or all) of vector $\mathbf{Z}_D$'s bits to 1. This results in many collisions during the scheduling phase. According to our experimental results, using the SR sub-protocol, the maximum number of SR restarts is 2, which can be considered as a threshold to decide on a DoS attack. When a DoS attack is detected during the scheduling phase, the DM sub-protocol is invoked which outputs the identity of disruptor(s). Then, after u_D is excluded from the list of peers, the honest users can resume the protocol.

Collusions: Consider N_C colluding users ($u_{c,1}, u_{c,2}, \ldots, u_{c,N_C}$) who want to deanonymize the messages of a specific user u_v. For this reason, they join the group of which u_v is a member, such that the final group size is N ($N > N_C$). Moreover, the N_C colluding users share their matrix of pairwise keys (i.e., $\mathbf{K}_i^{(0)}$) along with their random vector R_i. To deanonymize u_v's messages in round p, they need to compute u_v's XK vector $\mathbf{X}_v^{(p)}$, compare it with the received $\mathbf{Y}_v^{(p)}$, and obtain $\mathbf{m}_v^{(p)}$. They can compute $A_1(w) = \oplus_{j=1}^{N_C} \mathbf{k}_{v,j}^{(p)}(w)$ ($w = 1, 2, \ldots, N$) since they have already shared their matrix of pairwise keys and random vector R. Thus, they start to build $\mathbf{X}_v^{(p)}$:

$$\mathbf{x}_v^{(p)}(w) = \mathbf{A}_1(w) \oplus \mathbf{A}_2(w), \quad w = 1, 2, \ldots, N, \tag{4.50}$$

where $\mathbf{A}_2(w) = \bigoplus_{j=N_C+1}^{N} \mathbf{k}_{v,j}^{(p)}(w)$.

As you see, they need to have $A_2(w)$ to obtain each $\mathbf{x}_v^{(p)}(w)$ for $w = 1, 2, \ldots, N$. However, computing $A_2(w)$ requires the knowledge of pairwise keys $\mathbf{k}_{v,j}^{(p)}$ ($j = N_C + 1, N_C + 2, \ldots, N$) shared between u_v and the non-colluding users (u_{N_C+1}, $u_{N_C+2}, \ldots, u_N$). Hence, the attack is defeated since $A_2(w)$ is unknown to the colluding users. Furthermore, by employing a group entry control mechanism (like the one proposed in [85]), we can prevent malicious users from setting up large size collusion groups.

Node Failures: Suppose user u_o becomes offline while the protocol is being performed. This prevents the other users from computing SMV because u_o is no longer broadcasting his/her vector $\mathbf{Y}_o$. In this case, the remaining users can easily exclude u_o and resume the protocol. To do this, assuming u_o is disconnected at round p, every user u_i needs to exclude his/her keys shared with u_o (i.e. $\{\mathbf{k}_{io}^{(p)}(w)\}_{\substack{w=1 \\ w \neq i}}^{N}$) from Equations 10.1 and 10.2 before computing his/her $\mathbf{Y}_i$ vector. In other words, the users must remove the row associated with u_o from their matrix of pairwise symmetric keys $\mathbf{K}_i^{(p)}$ to be able to perform the next rounds of the protocol. However, they do not need to perform the initialization phase and set up new pairwise keys as the previous keys are still valid.

4.3.7 Performance Evaluation

In this section, we evaluate the performance of HSDC-net and present the results of our prototype implementation. In our evaluations, we consider Twitter, Facebook Messenger, and Instagram. We conducted our experiments based on the maximum number of characters per message allowed in these applications. For Twitter, the maximum number of characters in a tweet has recently increased from 140 to 280 characters . However, this value is 500 and 2000 characters for Instagram and Facebook Messenger, respectively . We considered 2 bytes of data per character on average, as UTF-8 coding system is used by these applications.

We developed a prototype implementation of HSDC-net to evaluate its deployment in microblogging applications. The implementation is written in *Python* and uses OpenSSL 1.1.0 for elliptic curve DSA signatures and PKI operations. We used the Deterlab [96] infrastructure to test the prototype under realistic network conditions. Deterlab provides a shared testbed for distributed protocols and enables us to easily change network topology and node configurations.

Setup: The testbed topology that we used in Deterlab consists of three 100 Mbps LANs with 10 ms latency between the core switches and clients. The three LANs are connected together using 10 Mbps links with 20 ms latency. We executed the protocol for 5 to 50 clients. Two types of client machines were used for the experiments: 3GHz Intel Xeon Dual Core with 2GB of RAM and 2.13 GHz Intel Xeon CPU X3210 Quad Core with 4GB of RAM (Figs. 4.18, 4.19 and 4.20).

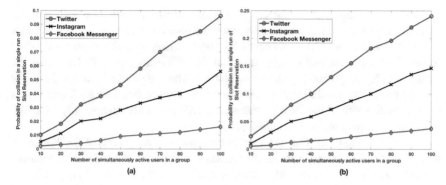

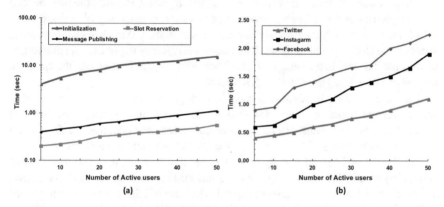

Fig. 4.18 Probability of collision after a single run of SR for **a** $B = 3$ and **b** $B = 5$

Fig. 4.19 **a** Time required to initialize the protocol, reserve a slot, and perform one cycle of anonymous message publishing. **b** End-to-end latency to publish an anonymous post

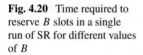

Fig. 4.20 Time required to reserve B slots in a single run of SR for different values of B

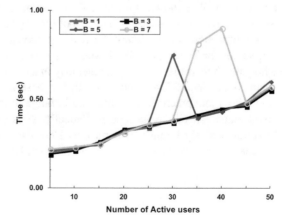

Collisions: Figures show the probability of collision for different values of the number of simultaneously active users and the scheduling overhead efficiency factor B. Collisions are more likely to occur for larger values of B that shows a sensible trade-off between collision probability and the efficiency factor B. Note that the values of collision probability shown in the figures have been obtained based on only a single run of the scheduling phase. However, we noticed that even for larger values of B, almost 100% of the slots for the next B MP cycles are successfully allocated in at most two runs of the scheduling phase.

Latency: It shows the time required to perform the three phases of HSDC-net. In this figure, the shown results are for the scenario in which the clients publish messages of length 560 bytes (the maximum size of a single tweet on average). Large values of N result in larger XK vectors that make the system slower. Note that the illustrated time required for performing the message publishing phase includes the time needed for a single run of the SR sub-protocol. For example, considering $N = 50$, it takes 1.1 s for the clients to anonymously publish a tweet in the group. 0.56 s of this time is spent on the slot reservation phase. The end-to-end latency to publish an anonymous post is for Twitter, Instagram, and Facebook Messenger. Twitter has the quickest responses since it has the smallest XK vector.

The time required to reserve B slots in a single run of the SR sub-protocol is shown. As the figure indicates, for $B = 1$ and $B = 3$ only a single run of SR is required to reserve B slots. However, some collisions have occurred for $B = 5$ and $B = 7$ in larger values of N that caused the SR sub-protocol to be restarted.

Finally, the end-to-end latency of HSDC-net and some of the most well-known anonymity protocols are compared. As you see, HSDC-net outperforms Dissent [84] and Verdict [80] protocols in terms of speed. The reason is that the SR and MP phases in HSDC-net are performed using simple and lightweight operations i.e. XOR and SUM. However, in Dissent and Verdict, heavy-duty tasks are performed for public-key encryption/decryptions and zero-knowledge proofs. Note that the latency of HSDC-net will be shorter if we have $B > 1$ because in this case only a single run of SR is required for B consecutive cycles of message publishing (Fig. 4.21).

Communication overhead: We also examined the maximum possible size of the XK vector for Twitter, Facebook Messenger, and Instagram. Considering 50 users who publish their messages simultaneously in a group, the maximum size of an XK vector is obtained 27, 49, and 195 KB, for these applications, respectively. These values are quite practical since they are in range of the average size of an ordinary email that is 75 KB. Note that, realistically, N is the number of users that want to simultaneously publish their messages not the maximum number of a group's members. Thus, the real group size can be much larger than N.

Fig. 4.21 End-to-end latency to anonymously publish a tweet for HSDC-net and some well-known anonymity schemes

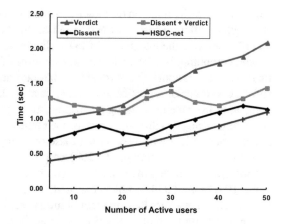

4.4 Personalized Privacy in Smart Homes

The proliferation of smart devices in recent years has led to novel smart home applications that upgrade traditional home appliances to intelligent units and automatically adapt their services without human assistance. In a smart home system, a central gateway is required to coordinate the functions of various smart home devices and allow bidirectional communications. However, the gateway may cause leakage of sensitive information unless proper privacy protections are applied. In this work, we first introduce a smart home model based on fog computing and secured by differential privacy. Then, we apply a personalized differential privacy scheme to provide privacy protection. Furthermore, we consider a collusion attack and propose our differential privacy model called APDP based on a modified Laplace mechanism and a Markov process to strengthen privacy protection, thus resisting the attack. Lastly, we perform extensive experiments based on the real-world datasets to evaluate the proposed APDP model.

4.4.1 Literature Review

The concept of a smart home has been widely explored in recent years [97, 98]. Alam et al. [99] describe the definition of a smart home as "an application of ubiquitous or pervasive computing or environment" and analyze the development of smart homes. Stojkoska et al. [100] present a holistic approach to the integration of state-of-the-art IoT solutions into smart homes. Chan et al. [101] present an international selection of leading smart home projects, as well as the associated technologies of wearable or implantable monitoring systems and assistive robotics. Datta et al. [102] develop an IoT architecture that enables smarter, connected and personalized healthcare and wellness services for residents of smart homes. Cicirelli et al. [103] propose a framework that primarily relies on the cloud-assisted agent-based smart home environment architecture, offering basic abstraction entities for design and implementation. Jie et al. [104] describe the integration of IoT technologies into smart home systems.

Fog computing has many advantages in terms of privacy protection and performance in a smart home. Dastjerdi et al. [105] introduce fog computing components, software systems and applications. Luan et al. [106] provide an overview of fog computing from the networking perspective to improve the efficiency aspects of fog computing. Chiang et al. [107] describe the range of new challenges in the emerging IoT and the difficulty of overcoming these challenges with today's computing and networking models. Brogi et al. [108] propose a general and extensible model to support QoS-aware deployments of IoT applications in a fog infrastructure. Bo Tang et al. [109] present a hierarchical distributed fog computing architecture to support the integration of a very large number of infrastructure components and services into future smart cities. Soumya Kanti Datta et al. [110] discuss the architecture of fog computing that is deployed at roadside units (RSUs) and M2M gateways that offers consumer-centric IoT services. Wangbong Lee et al. [111] present a gateway-based fog computing architecture for wireless sensor and actuator networks (WSANs).

Security and privacy issues in smart homes have been extensively considered by many researchers. Several existing approaches provide privacy protection uniformly. The clustering-based methods include K-anonymity [14], L-diversity [15], T-closeness [16] and their variants [112]. Clustering-based methods provide satisfactory protection under the scenario of datasets with records that share the same attributes; however, such methods do not work well in the data diffusion scenario. Dwork proposed differential privacy that offers privacy protection with a solid theoretical foundation [17]. Under the framework of differential privacy, numerous mechanisms have been proposed to achieve privacy preservation, e.g., Laplace noise [113] and sampling [40]. Although differential privacy can offer strict protection, most existing approaches use fixed privacy levels to cope with various requirements. Recent privacy studies of smart homes also include the communication protocol [114], data analytics in cloud-based smart homes [115], location sharing [116], etc.

In this era of big data, privacy protection is required in every aspect of the system [117, 118]. Komninos et al. [119] present dangers encountered in some of the most illustrative scenarios of interaction among entities of the smart home and smart grid environments, evaluating their impact on the entire grid. Geneiatakis et al. [120] set up the scene for a security and privacy threat analysis for a typical smart home architecture that relies on existing IoT devices and platforms that are readily available in the market. Lee et al. [121] discuss the concept of the IoT fog as well as the existing security measures useful in securing the IoT fog and then explore potential threats to the IoT fog. Yang et al. propose privacy preserving collaborative filtering via the Johnson-Lindenstrauss transform [122]. Zhang et al. [123] introduce various aspects of smart city applications, discuss the system architecture, then present the general security and privacy requirements, and identify several security challenges for the smart city.

Personalized privacy can optimize the data utility while reducing the overall privacy budget [124, 125]. For personalized privacy, Wang et al. [4] use a Markov decision process to control the granularity of the published data. Koufogiannis et al. [61] leverage personalized differential privacy to protect the privacy based on distance in social networks. Götz et al. [126] explore privately releasing user context streams for personalized mobile applications. In addition, Aghasian et al. propose a method to measure the privacy closure based on multiple social networks [127]. However, all exisiting works barely consider the personalized privacy protection in smart home scenario. In addition, personalized privacy will result in collusion attack in a certain extent, which is merely discussed either. We will try to solve these issues in the rest of this work.

4.4.2 Smart Home Modeling Based on Fog Computing and Differential Privacy

The objective of the smart home model is to provide high-quality services to the user while maximizing the network bandwidth and minimizing processing latency. In the proposed model, we consider a fog structure and use a fog server as the bridge between the cloud server and the IoT applications. The fog server has a certain computing capability and storage used to perform computational tasks, reducing the processing time and service latency. For clarity, we first analyze the details of the smart home architecture and then model it as a graph based on graph theory. The rationale is that we regard fog servers, cloud servers, and smart devices as nodes and the connections between them as edges.

4.4.2.1 Fog Computing-Based Smart Home Model

To offer quality services at home, a smart home can provide many different components. In this section, we discuss the architecture of general application with basic components.

Figure 4.22 demonstrates the architecture of a smart home based on fog computing, which has four layers: IoT devices, the fog server, the public cloud, and the application layer. The IoT layer includes smart devices deployed at home, such as sensors, wearable devices, smart meters, electric devices, and monitoring devices. With these smart devices, the IoT layer can obtain all of the status information in a smart home and send it to the fog simultaneously or perform post-processing at the fog server. Therefore, the IoT layer primarily performs the functions of data collection and service provision to users. The fog layer has computing, storage, control, communication, security, and privacy protection capabilities. It can process most of the data collected from the IoT and sends the analytic results to the cloud or provides direct feedback to IoT devices. When the data size is beyond the fog's processing ability, a request is sent to the cloud to participate in further processing. This fog-based structure can improve the real-time processing ability in a smart home, reduce

Fig. 4.22 Hierarchical structure of a smart home

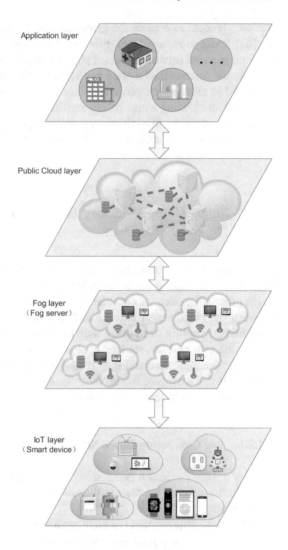

the system latency, and save the network bandwidth. The public cloud layer has superior computing and storage capabilities that provide support to the fog. In addition, it can provide a variety of access types to the application layer, which is the top layer. Entities provide services such as medical center, alarm center, and electric utility.

Building on the fog-based smart home structure, we model a smart home as a graph. We develop a personalized differential privacy protection model. It can minimize the overall privacy budget while improving data utility. In Sect. 4.4.5, we leverage a modified Laplace mechanism that introduces a noise generation process into a stochastic process and decouples the correlation among noises. As a result, we can eliminate the collusion attack under this scenario.

4.4.2.2 Trust Distance-Based Differential Privacy

Based on the proposed smart home structure, we further model it using graph theory. We use a weighted graph

$$G = \left\{n, e, w \middle| n \in N, e \in E, w \in W\right\} \tag{4.51}$$

to denote the smart home based on the fog computing paradigm. In graph G, we use $n \in N$ to represent each node, $e \in E$ to denote the relationship between nodes, and $w \in W$ to show the weights between nodes. If there are two nodes n_i and n_i and at least one series of edges $\{e_{ik_1}, e_{ik_2}, \ldots, e_{k_n j}\}$ connects them, we conclude that the nodes have a relationship. Based on the relationship, we also use d^T to describe the trust distance, where $d^T \in D^T$ t.

The application layer contains many nodes (applications), such as TVs, lighting, and cyber-physical equipment. The nodes may further connect to several sub-nodes. In this case, the nodes in the end are specific to a certain function and may leak the user's private information, e.g., the blood pressure measurement. Based on this observation, we set the trust distance D^T as the number of hops between the nodes and the fog server. Therefore, the privacy level ϵ increases with the growth of trust distance D^T, while ϵ decreases with the reduction of D^T.

For each node n_i, when fog server n_F tries to diffuse its data d_{ij} to cloud n_{Cj}, the fog server generates a proxy of the data as

$$\hat{y}_{ij} = d_{ij} + Lap\left(\frac{\delta}{\epsilon}\right), \tag{4.52}$$

where $Lap(\frac{\delta}{\epsilon})$ denotes the Laplace noise [113], while δ and ϵ are the global sensitivity and privacy level, respectively.

The node n_i demands that the proxy $\hat{y}_{ij}$ satisfy $\epsilon(D^T)$-differential privacy to protect the private data. For privacy level $\epsilon(D^T)$, D^T is a distance function denoting the distance between node n_i and fog server n_F.

For simplicity, we regard graph G as an undirected graph. However, this assumption can be eliminated, as the model works the same way for directed graphs. We also assume that the fog server is a trusted central authority and can process the data with ϵ-differential privacy and transmit the data via a secure communication channel. In the fog server, the privacy budget is a constant B that equals the sum of all privacy levels of all the published data.

4.4.3 Personalized Differential Privacy Scheme

We propose a personalized differential privacy-preserving data publishing model of a smart home, where the sensitive data of a certain node n_i may be shared with

Table 4.3 Sensitive data classification

Type	Contents
Location	$d_i \in R_3$ represents the GPS coordinates: longitude, latitude, and elevation
Timestamp	$d_i \in R_+$ indicates a positive instance of time, e.g., "2000-12-20 00:00:00"
Dual states	$d_i \in \{0, 1\}$ indicates a binary status such as the standby status
Text data	$d_i \in ABC$ denotes a string of alphabetic characters like the name of a TV series

the cloud through fog server n_F. The usual data types are listed in Table 4.3. The sensitive data's privacy level usually varies with and is based on the trust distance. The reason is that the contents are more private and specific if the trust distance is longer. The nodes with the longest trust distance, e.g., the wearable equipment and smart meters, possess the most private data. Moreover, the sensitive data may be released to multiple clouds if the resources of a single cloud cannot satisfy the needs. For multiple clouds, the sensitive data should be provided different levels of protection according to various requirements. In both cases, the fog server should provide personalized privacy protection to the sensitive data of node n_i.

4.4.3.1 Differential Privacy

Differential privacy has a solid theoretical foundation for providing privacy protection to two adjacent datasets. In two adjacent datasets denoted by D and D', it is required that D' have one more record than D and that an adversary cannot re-identify this specific record.

ϵ is a positive privacy parameter decided by the overall privacy budget of the system. D' and D are two adjacent datasets with an adjacent relationship. Denote by $\mathcal{A}$ a randomized algorithm that sanitizes the datasets. The algorithm M is called ϵ-differentially private on D' and D if and only if

$$\Pr\left[\mathcal{A}(D') \in \Omega\right] = \exp\left(\epsilon\right) \times \Pr\left[\mathcal{A}(D) \in \Omega\right], \qquad (4.53)$$

where the probability space Ω is taken over the randomness used by $\mathcal{A}$.

4.4.3.2 Laplace Mechanism

The Laplace mechanism is most typically used to attain ϵ-differential privacy in a numeric scenario. The key feature of this method is the generation of a random noise that follows the Laplace distribution. After the noise has been added to the raw data, an adversary cannot re-identify the location of the extra record.

The mechanism $\mathcal{M} : R^n \rightarrow \Delta(R^n)$ that adds Laplace-distributed noise $\mathcal{N}$ is defined by

$$\mathcal{M}(\mathcal{D}) = \mathcal{D} + \mathcal{N},$$

s.t.

$$N \sim Lap\left(\frac{\delta}{\epsilon}\right), \tag{4.54}$$

$$Lap\left(b\right) \sim d\Pr\left[N = n\right] = \exp\left(-\frac{||n||_2}{b}\right),$$

where $d\Pr[N = n]$ is the density of $Lap(b)$. Following the above formulation, we regard $\mathcal{M}$ as an ϵ-differentially private mechanism under an adjacency relation.

4.4.3.3 Privacy Protection Based on Trust Distance

Building on the smart home model provided in Sect. 4.4.2.2, we formulate the personalized privacy protection model based on trust distance in a fog computing structure.

In a smart home, if the computing power of the fog server is insufficient, the sensitive data of node n_i may be passed to cloud server n_{Cj}. Such data need to be protected for privacy reasons. Most existing approaches usually consider uniform privacy level protection and apply the protection once and for all. However, uniform protection cannot meet the new requirements of the smart home for various sources of data and possible multiple clouds. Therefore, we focus on providing personalized privacy protection to smart homes in this section.

In this section, we focus on personalizing the trust distance D_T to avoid private data leakage in a single cloud. The objective is to design a differential privacy mechanism $\{A : D \rightarrow \delta D_n\}$ that publishes the sensitive data d_i from node n_i to recipient node n_j. The mechanism A generates n outcomes $\hat{y}_{ij}^n$ that are then released to the cloud. Furthermore, A needs to meet the following constraints.

The first constraint is providing personalized privacy protection. For all the sensitive data d_{ij}^T, the generated n proxies $\hat{y}_{ij}^n$ have to satisfy $\epsilon(d_{ij}^T)$-differential privacy, where $MAP()$ is a mapping function that maps trust distance d_{ij}^T to privacy level ϵ.

The second constraint is limiting the upper bound of personalized privacy levels after composition. For all the proxies $\hat{y}_{ij}^n$, the ceiling of all composition mechanisms should equal the maximum $\epsilon(d_{ij}^T)$, rather than the sum of all $\epsilon(d_{ij}^T)$.

$$\sum_{i=1;j\neq i}^{n} \mathcal{A}_{DP}\left(MAP\left(d_{ij}^T\right)\right) = \max \mathcal{A}_{DP}\left(MAP\left(d_{ij}^T\right)\right), \tag{4.55}$$

where the superscript DP denotes differential privacy.

The third constraint is to obtain the maximum utility under the personalized privacy scenario. For all the noisy responses $\hat{y}_{ij}^n$, they should denote the most accurate outputs of the raw data d_{ij}. The least noisy response results in the maximum data utility. In terms of numeric value, the data utility is usually measured by the root-mean-square error. Thus, the minimum root-mean-square error results in the maximum data utility.

In the proposed personalized privacy protection model, there are various noisy outputs $\hat{y}_{ij}^n$, and therefore, there are multiple corresponding data utility values. In this work, we specifically refer to the sum of data utility values when considering the maximum data utility in Eq. 4.56.

$$\sum_{i=1, j\neq i}^{n} \mathrm{E}\left\|\hat{y}_{ij}^n - d_{ij}\right\|_2^2. \tag{4.56}$$

4.4.3.4 Privacy Protection in Multiple Clouds

In addition to the features of the sensitive data itself, the data may also be released to multiple clouds from the fog server. The privacy challenges have long been discussed by earlier studies. In this subsection, we try to solve the personalized privacy protection problem in the multiple-cloud scenario. For instance, when the TV station asks for a TV series, the fog server will pass the query to a certain cloud server n_{Cj}. However, after a few episodes, the fog server may observe that this cloud has stopped storing the TV series, or lacks a few episodes. In this case, the fog server has to send the query to other clouds to obtain assistance. In this way, the sensitive data are released to multiple clouds, and privacy protection is necessary.

The fog server publishes node n_i's sensitive data $k \in K$ times, where K denotes the number of clouds that receive the same data. The personalized privacy level is represented by $MAP(k)$. It is a mapping function that maps k to privacy level ϵ.

We consider a one-round relaxation example for clarity. The results can also be intuitively extended to multiple rounds.

Assume that there are two privacy levels $\epsilon(K)$ and $\epsilon(K')$, where $\epsilon(K') > \epsilon(K)$. There is a mechanism $\mathcal{A}_{\epsilon(K)\to\epsilon(K')} : \mathcal{D} \to \Delta(\hat{\mathcal{Y}}^2)$ that publishes the sensitive data to two different cloud servers. In the first cloud server, node n_i publishes a proxy $\hat{y}_{ij}^K$ to cloud server n_{Cj}. The proxy $\hat{y}_{ij}^1$ satisfies $\epsilon(K)$-differential privacy. In the next cloud server, the privacy level is relaxed to $\epsilon(K')$-differential privacy. If clouds collude to steal the more accurate sensitive data, the proposed mechanism should satisfy

$$\mathcal{A}_{DP}\Big(\epsilon(K) + \epsilon(K')'\Big) = \mathcal{A}_{DP}\Big(\epsilon(K')\Big), \tag{4.57}$$

where $\epsilon(K')'$ is the privacy level of the second noisy response. As the upper bound of the composition theorem indicates, we have

$$\mathcal{A}_{DP}\Big(\epsilon(K')'\Big) = \mathcal{A}_{DP}\Big(\epsilon(K') - \epsilon(K)\Big), \qquad (4.58)$$

from which we can obtain $\epsilon(K')' < \epsilon(K')$. This result implies that the second proxy cannot relax the privacy level at all but must instead tighten the privacy level. This is a contradiction, especially if $\epsilon(K) < \epsilon(K') \ll 1$. The data utility degrades significantly, resulting in applications becoming impractical. The problem of personalized privacy-preserving data publishing in multiple clouds can be formulated as described below.

We propose a mechanism $\{A_{\epsilon(1) \to \epsilon(K)} : D \to \delta \hat{y}^k\}$ that is differentially private if the sensitive data are published in multiple clouds. The mechanism A generates multiple proxies $\hat{y}^k$ and releases them to k different clouds. With the increase of $\epsilon(K)$, the outcomes become progressively more accurate. In this scenario, the mechanism A should further satisfy the following constraints.

The first constraint is providing personalized privacy protection. For all the sensitive data d_{ij}^T, the generated n proxies $\hat{y}_{ij}^n$ have to satisfy $\epsilon(d_{ij}^T)$-differential privacy, where $MAP()$ is a mapping function that maps trust distance d_{ij}^T to privacy level ϵ.

The second constraint is limiting the upper bound of personalized privacy levels after composition. For all the proxies $\hat{y}_{ij}^n$, the ceiling of all composition mechanisms should equal the maximum $\epsilon(d_{ij}^T)$, rather than the sum of all $\epsilon(d_{ij}^T)$.

$$\sum_{i=1; j \neq i}^{n} \mathcal{A}_{DP}\Big(MAP\big(d_{ij}^T\big)\Big) = \max \mathcal{A}_{DP}\Big(MAP\big(d_{ij}^T\big)\Big), \qquad (4.59)$$

where the superscript DP denotes differential privacy.

The third constraint is to obtain the maximum utility under the personalized privacy scenario. For all the noisy responses $\hat{y}_{ij}^n$, they should denote the most accurate outputs of the raw data d_{ij}. The least noisy response results in the maximum data utility. In terms of numeric values, data utility is usually measured by the root-mean-square error. Thus, minimizing the root-mean-square error results in the maximum data utility.

The last requirement is personalized privacy levels of data publishing in multiple clouds. For data release to multiple clouds, the privacy levels $\{\epsilon(1), \epsilon(2), \ldots, \epsilon(K)\}$ increase monotonically, which can be described by $\epsilon(1) < \epsilon(2) < \cdots < \epsilon(K)\}$.

4.4.3.5 Generic Personalized Privacy Scheme for a Smart Home

In the above subsections, we analyze two types of personalized privacy protection application scenarios. The results show that personalized differential privacy is necessary, especially in a fog server. As the computing power of a fog server is limited, the privacy budget provided is also limited and fixed in a relevant small range.

For the trust distance problem and the multiple-cloud problem, we observe that both involve the same challenge, i.e., personalized privacy level functions $\epsilon()$. Therefore, they are the same problem in a certain sense. Building on this, we formulate the problem of trust distance-based personalized privacy data diffusion over multiple clouds as follows.

Our target is to design a differentially private mechanism $\{A_{\epsilon(d_{i1}^T,1)\to\epsilon(d_{ij}^T,k)} : D \to \delta\hat{y}^T\}$ that publishes the sensitive data from node n_i to cloud server n_{Cj} through the fog server in a smart home. The mechanism generates $n \times k$ noisy proxies that are published to k cloud servers. The privacy level $\epsilon(d_{ij}^T, k)$ increases with incremented k, and the noisy proxies become progressively more accurate. The privacy level $\epsilon(d_{ij}^T, k)$ decreases as d_{ij}^T increases, and the proxies become progressively more noisy. Additionally, the mechanism A needs to satisfy the following constraints.

- Multiple-cloud data release: For a data release to multiple clouds with a fixed distance d_{ij}^T, the privacy levels $\{\epsilon(d_{ij}^T, 1), \epsilon(d_{ij}^T, 2), \dots, \epsilon(d_{ij}^T, T)\}$ increase monotonically, as represented by $\epsilon(d_{ij}^T, 1) < \epsilon(d_{ij}^T, 2) < \cdots < \epsilon(d_{ij}^T, T)$.
- Personalized privacy protection: For all the sensitive data d_{ij}, each proxy $\hat{y}_{ij}^T$ should satisfy $\epsilon(d^T(ij), k)$-differential privacy.
- Limited upper bound of composition: For all the noisy proxies y_{ij}^k, the ceiling after composition should be the maximum $\epsilon(\frac{1}{d_{ij}}, t)$, rather than the sum of $\epsilon(\frac{1}{d_{ij}}, t)$. The mathematical description is provided by Eq. 4.60.

$$\sum_{i=1, j\neq i, K}^{n,n,K} \mathcal{A}_{DP}\left(MAP\left(d_{ij}^T, L\right)\right) = \max_{d_{ij},k} \mathcal{A}_{DP}\left(MAP\left(d_{ij}^T, k\right)\right). \quad (4.60)$$

- Maximum data utility: All the proxies y_{ij}^T have to be the most accurate noisy responses of actual outputs d_{ij}, which results in the maximum utility. In the numeric data sense, data utility is measured by the root-mean-square error, as shown in Eq. 4.61. Furthermore, the minimum root-mean-square error leads to the maximum data utility.

$$\sum_{i=1, j\neq i, k}^{n,n,K} E\left\|y_{ij}^T - d_i^T\right\|_2^2. \quad (4.61)$$

4.4.4 Collusion Attack Under Differential Privacy

After analyzing the two personalized privacy protection scenarios, we observe that adversaries can launch collusion attacks that impact the protection's effectiveness. Under the personalized differential privacy protection scenario, adversary and collusion attacks have certain new features and can be formulated mathematically under differential privacy.

We model the adversary in the differential privacy sense in this work. In most existing approaches, the adversary is considered qualitatively, rather than quantitatively. Furthermore, the attack cannot be formulated based on the adversary. The result is that we can only measure the relative attack impact, e.g., via the information theory-based entropy. Therefore, we propose the differential privacy-based adversary model as follows.

In the proposed personalized differential privacy model, the privacy levels are modeled by $\epsilon(\cdot)$. Therefore, we model the adversary by

$$\mathcal{P}_{ad} = \mathcal{A}_{DP}\left(\epsilon_{ad}\right), \tag{4.62}$$

where the background knowledge of the adversary can be regarded as complying with ϵ_{ad}-differential privacy.

The advantage of modeling the adversary in this way is that we can use the composition theorem to include the impact of the adversary in the privacy protection model. In addition, the collusion attack can be further analyzed based on this definition.

A collusion attack is widely known as the scenario of two or more adversaries colluding with each other to obtain more accurate data. In our case, two or more clouds may share their data to perform a collusion attack and cause a leakage of private data. There are three conditions for launching a collusion attack. First, the sensitive data have been published on two or more clouds. Second, each of the clouds already possesses some data, and the clouds share the same interest. Third, the clouds have the incentive that they can obtain more information after colluding.

Building upon the adversary model, we can further develop the collusion attack definition.

Given $m \in M$ adversaries (cloud servers) and their corresponding privacy levels $\epsilon d^T, k_M$, the collusion attack can be described by

$$\mathcal{CA}(D^T, K, M) = \sum_{d^T \in D^T, k \in K, m \in M}^{D^T, K, M} \epsilon(d^T, k)_m \tag{4.63}$$
$$= \epsilon(d^T, k)_1 + \epsilon(d^T, k)_2 + \cdots + \epsilon(d^T, k)_m,$$

where $\mathcal{CA}(\cdot)$ is the sum of all privacy levels (ϵs). As discussed above, the increase of the privacy level ϵ leads to a degradation of privacy protection. As the composition theorem is a built-in feature of differential privacy, collusion attacks can always be launched without proper operation.

4.4.5 APDP Model

In our proposed smart home model, the privacy protection is guaranteed by differential privacy with Laplace noise. However, the existence of composition features of differential privacy may result in degradation of privacy protection. Therefore, we introduce the APDP model that uses a modified Laplace mechanism, in which the noise generation is integrated with the Markov process. As a result, the correlations among noises are broken, and hence, our APDP model is able to resist the collusion attack.

4.4.5.1   Composition Mechanism Underlying APDP

APDP is created to incorporate various mechanisms to provide privacy protection. In addition to the Laplace mechanism, other mechanisms include the exponential mechanism, the Gaussian mechanism, sampling, etc.

Assume that mechanisms $\{\mathcal{A}_1, \mathcal{A}_2, \ldots, \mathcal{A}_n\} : \mathcal{D} \to \Delta(\mathcal{Y})$ respectively satisfy $\{\epsilon_1, \epsilon_2, \ldots, \epsilon_n\}$-differential privacy. The composition mechanism $\mathcal{A} : \mathcal{D} \to \Delta(\mathcal{Y}^n)$ defined by $\mathcal{A} = \{\mathcal{A}_1, \mathcal{A}_2, \ldots, \mathcal{A}_n\}$ is called $\sum_i^n \epsilon_i$-differentially private.

$$\mathcal{A}_{DP}\left(\epsilon_{com}\right) = \sum_i^n \mathcal{A}_{DP}\left(\epsilon_i\right). \tag{4.64}$$

The respective privacy level $\sum_i^n \epsilon_i$ denotes the upper bound of the composition theorem. However, $\sum_i^n \epsilon_i$ overstates the actual privacy level. In this section, we introduce APDP to explore correlations among mechanisms that ensure better privacy protection.

4.4.5.2   Incorporating a Markov Process

I have modified the section to highlight APDP; please change accordingly to suit the change.

For n-dimensional numeric data d, our target is to propose a smart home-suitable mechanism A inside APDP to generate the noisy outputs $\hat{y}_{ij}^K$ that are sent by smart home node n_i to cloud server n_j and published in K cloud servers. A must have two features. First, the accuracy $||\hat{y}_{ij}^K - d_{ij}^K||$ should solely depend on the trust distance d^T and the number of multiple cloud servers k, while all the other responses do not degrade the accuracy. Second, any group of cloud servers has no ability to infer more sensitive information about smart home node n_i after collusion $\sum \epsilon(d_T, k)$.

Motivated by this, we introduce a Markov process inside APDP that is defined over a continuous domain. This Markov transfer process will be further applied to fulfill the privacy-preserving mechanism in the following study.

Given the privacy level ϵ and three specific privacy levels ϵ_{i-1}, ϵ_i, and ϵ_{i+1}, where $\epsilon_{i-1} < \epsilon_i < \epsilon_{i+1}$, the Markov process has the following properties.

- The noise follows the Laplace distribution: $\forall \epsilon > 0, d\Pr\left(V_\epsilon = v\right) \propto \exp\left(-\epsilon||v||_2\right)$.
- The noise generation process is a Markov process: $\forall \epsilon_{i-1} < \epsilon_i < \epsilon_{i+1}, V_{\epsilon_{i-1}} | V_{\epsilon_i}, V_{\epsilon_{i-1}} \perp V_{\epsilon_{i+1}}$.

- The transfer probability of the Markov process is

$$d \Pr \left(V_{\epsilon_i} = v_i \middle| V_{\epsilon_{i+1}} = v_{i+1} \right) \propto \delta(v_i - vi + 1)$$

$$+ \frac{(n+1)\epsilon_i^{1+\frac{n}{2}} ||v_i - v_{i+1}||_2^{1-\frac{n}{2}}}{(2\pi)^{\frac{n}{2}}} B_{\frac{n}{2}-1} \left(\epsilon_i ||v_i - v_{i+1}||_2 \right) \tau$$

$$+ O\left(\tau^2 \right) \tag{4.65}$$

s.t.

$$\tau = \frac{\epsilon_i}{\epsilon_{i+1}} - 1,$$

where B is the Bessel function.

We need the Markov process to guarantee that the correlations between noises are properly decoupled. Therefore, the proposed APDP model is able to resist the collusion attack (Fig. 4.23).

4.4.5.3 APDP Analysis

The Laplace mechanism is a popular approach to satisfying ϵ-differential privacy requirements. However, it cannot be optimal in terms of the minimum mean-square error. Therefore, in APDP, we target achieving the optimum Laplace mechanism for both minimum entropy and minimum mean-square error by designing the noise properly. (You need to play down the tone of the mechanism while highlighting APDP, which has a built-in mechanism and is customizable and configurable.)

Given the ϵ-differentially private mechanism $A : R^n \to \Delta(R^n)$, A satisfies $y_{ij}^K = d_{ij} + N$, where $N \sim \rho(N) \in \Delta(R^n)$. The mean-square error can be minimized if the noise density f satisfies

$$f_1^n(v) = \left(\frac{\epsilon}{2} \right) \exp \left(-\epsilon ||v||_1 \right), \tag{4.66}$$

where $f_1^n(v)$ denotes the density of noise at v. Thus, we have

$$E \left\| y_{ij}^t - d_{ij} \right\|_2^2 = E_{V \sim \rho} \left\| V \right\|^2$$

$$\geq E_{V \sim f_1^n} \left\| V \right\|_2^2 \tag{4.67}$$

$$= \frac{2n}{\epsilon^2}.$$

The optimum Laplace mechanism provides the solution to achieving optimized data utility at a fixed privacy level. We further prove that the proposed method can

Fig. 4.23 Privacy level
comparison in multiple
clouds

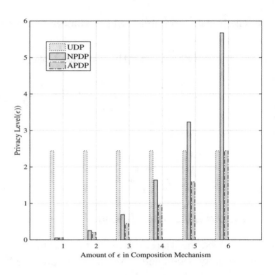

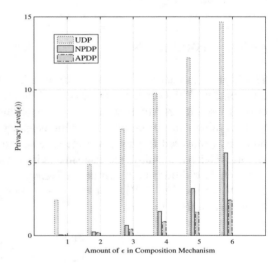

satisfy the optimum Laplace mechanism, which makes the proposed model more
feasible and practical.

First, a one-dimensional case is considered for clarity. It can be further extended
to multiple dimensions. In the following proposed theorem, we establish a method
that satisfies all the requirements and illustrates the feasibility and effectiveness.

Two privacy levels, ϵ_1 and ϵ_2, which represent abbreviated notations of $\epsilon_1\left(\frac{1}{d_{ij}}, t\right)$
and $\epsilon_2\left(\frac{1}{d_{ij}}, t\right)$, satisfying $0 < \epsilon_1\left(\frac{1}{d_{ij}}, t\right) < \epsilon_2\left(\frac{1}{d_{ij}}, t\right)$, are given. Then, the form of
the mechanism is

$$y_{i1}^t = d + V_1, \; y_{i2}^t = d + V_2, \; (V_1, V_2) \sim \rho\Delta(R^2). \tag{4.68}$$

Moreover, the density $f_{\epsilon_1(\frac{1}{d_{ij}},t),\epsilon_2(\frac{1}{d_{ij}},t)}$ is

$$\begin{aligned}
f_{\epsilon_1,\epsilon_2}(x, y) = {} & \frac{\epsilon_1^2}{2\epsilon_2} \exp\left(-\epsilon_2|y|\right)\delta(x - y) \\
& + \frac{\epsilon_1(\epsilon_2^2 - \epsilon_1^2)}{4\epsilon_2} \exp\left(-\epsilon_1|x - y| - \epsilon_2|y|\right).
\end{aligned} \tag{4.69}$$

Based on the theorem, we can conclude that the theorem has the following properties.

- The mechanism $\mathcal{A}_1$ is $\epsilon_1\left(d_T, k\right)$-differentially private.
- The mechanism $\mathcal{A}_1$ is optimal. Namely, $\mathcal{A}_1$ minimizes the mean-square error $E(V_1)^2$.
- The mechanism $\mathcal{A}_2$ is $\epsilon_2\left(d_T, k\right)$-differentially private.
- The mechanism $\mathcal{A}_2$ is optimal. Namely, $\mathcal{A}_2$ minimizes the mean-square error $E(V_2)^2$.
- The mechanism $(\mathcal{A}_1, \mathcal{A}_2)$ is $\epsilon_2\left(d_T, k\right)$-differentially private.

The rationale for the noise following a Markov stochastic process is that a Markov process requires that the current state be only related to the preceding state. This implies that the current state is not impacted by the other states before the preceding state. In this case, the current noise is only determined by the preceding noise. In the proposed model, the privacy level increases with the trust distance, as does the noise. Therefore, the current user has no incentive to collude with the next user who has an inaccurate output with a greater noise.

4.4.6 Performance Evaluation

In this part, we demonstrate the performance of the proposed model in terms of privacy protection and data utility and compare the proposal to the ordinary personalized differential privacy that only personalizes the privacy levels but does not consider resistance to attacks. We also show that our proposal can outperform the existing approaches from the perspective of background knowledge attack. As a result, the experimental results based on real-world datasets show that our proposed model can minimize the overall privacy budget and maximize the data utility while eliminating the background knowledge attack.

We use a real-world smart home dataset that is collected in [128, 129] and is based on a health-related smart home. The data are collected under 7 scenarios, including sleeping, resting, dressing, eating, toilet use, hygiene and communication. Fifteen candidates are contained in this dataset. Specifically, the shortest path is denoted by the hop account that captures the features of our model. However, our model can accommodate any type of distance metric used in existing approaches.

The algorithms are implemented in MATLAB 2015 and run on a Mac OS platform with a Core i5 CPU running at 2.7 GHz with 8 GB of RAM.

In the comparison experiments, we compare the proposed attack-proof personalized differential privacy model (APDP) with uniform differential privacy (UDP) [17] and ordinary or normal personalized differential privacy (NPDP) [61]. First, UDP provides uniform privacy levels to all nodes. Second, NPDP provides different privacy levels based on various requirements. Third, APDP offers attack-proof personalized privacy levels built upon NPDP. We demonstrate the evaluation results below.

4.4.6.1 Privacy Protection

As Fig. 4.24 shows, we use 1 to 6 nodes to simulate the composition mechanism. We can conclude that the proposed APDP has the best performance in term of privacy

Fig. 4.24 Privacy level comparison in multiple clouds against collusion attack

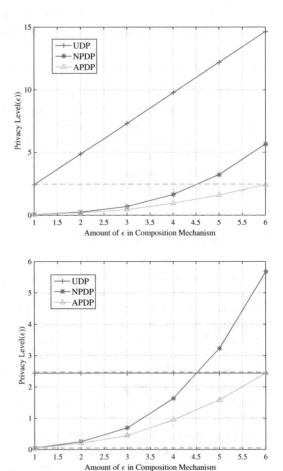

protection. With the increase of the node quantity, the privacy level of UDP does not increase and remains stable. The privacy levels of both NPDP and APDP increase due to the impact of the composition mechanism. Comparing these two models, we observe that APDP increases quickly and that the privacy levels release fast. The privacy issues are quite severe. However, APDP performs better, as it increases slowly, and the maximum values equals that of UDP. Therefore, APDP can minimize the negative impact of the composition mechanism and provide better privacy protection from the perspective of both strictness and customization (Fig. 4.24).

In Fig. 4.24, we illustrate the case of multiple clouds instead of multiple nodes. Similarly, we use 6 clouds as an example. In the case of multiple clouds, all the clouds are independent, and there is no co-relation inside them. Therefore, the composition theorem has a more significant impact on privacy protection. All three models suffer from performance degradation with the increase in the number of clouds. We can conclude that the privacy level of UDP increases the fastest, followed by NPDP. Only the privacy level of APDP increases moderately, and the maximum privacy level is still satisfactory.

In summary, APDP has the best privacy protection compared to UDP and NPDP under the scenarios of both multiple nodes and multiple clouds.

4.4.6.2 Data Utility

In the case of data utility, we reach the following conclusions based on Fig. 4.25. The vertical axis denotes the amount of injected noise. Therefore, the smaller the noise is, the higher the data utility. The trends show that the data utility of UDP maintains the same level and remains the highest regardless of the number of multiple clouds. As for NPDP and APDP, the utilities of both approaches increase with the number

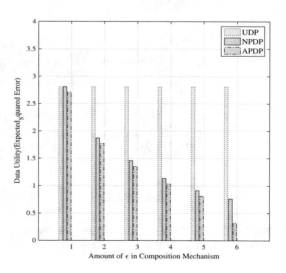

Fig. 4.25 Data utility comparison

of multiple clouds. However, compared to NPDP, APDP rises faster, i.e., it provides a higher data utility. In addition, the more clouds there are, the higher the data utility.

4.4.6.3 Defense Against a Collusion Attack

In Fig. 4.26, we illustrate the performance of the three models against a collusion attack in multiple clouds. In this case, there are two attacks, which are represented by two green dashed lines. We demonstrate that UDP cannot prevent a collusion attack, while NPDP and APDP have the ability to defeat the attack to different degrees. NPDP can resist an attack to a certain degree; however, it ultimately fails as the green dashed line has an intersection with the red line. However, APDP is fully attack-proof, as the yellow line is consistently under the green dashed line.

In Fig. 4.26, we illustrate the performance of the three models against a collusion attack with multiple nodes. Similar to the above, there are two attacks, which are represented by two green dashed lines. We demonstrate that UDP can prevent a collusion attack because the privacy level is not released after composition. NPDP and APDP have the ability to defeat the attack to different degrees. NPDP can resist the attack to a certain degree; however, it ultimately fails as the green dashed line has an intersection with the red line. APDP is fully attack-proof, as the yellow line is consistently under the green dashed line.

To summarize, APDP has the best performance in terms of collusion attack resistance. It can eliminate the collusion attack due to the properly decoupled noise generated by the modified Laplace mechanism.

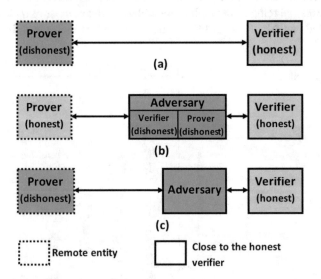

Fig. 4.26 Distance-bounding protocols are generally exposed to three types of security attacks: **a** Distance Fraud, **b** Mafia Fraud, and **c** Terrorist Fraud

4.5 Personalized Privacy in Location-Based Services

In recent years, advancements in smartphone technology and positioning systems have resulted in the emergence of location-based applications and services such as activity-tracking applications, location-based services (LBS), database-driven cognitive radio networks (CRNs), and location-based access control systems. In these services, mobile users' real-time location data is utilised by a location-based service provider (LBSP) to provide users with requested information or access to a resource or service. These applications are fast growing and very popular due to the range of useful services they offer [130, 131].

However, it is possible for dishonest users to submit fake check-ins by changing their GPS data. To clarify and highlight the fake location submission issue consider LBSPs like Yelp and Foursquare that may offer some rewards (such as gift vouchers) to users who frequently check in at specific locations. This creates an incentive for dishonest users to submit fake check-ins by manipulating their GPS data. For example, in a research study, Zhang et al. [132] found that 75% of Foursquare check-ins are false and submitted by dishonest users to obtain more rewards. Furthermore, in database-driven CRNs, malicious users can submit fake locations to the database to access channels which are not available in their location [133].

In this chapter, we highlight and review the existing location verification schemes. These schemes are also called location proof (LP) schemes in the literature. Moreover, we present some preliminaries as the foundation for the next three chapters.

4.5.1 Literature Review

In this section, we review the literature on location proof (LP) schemes. They are generally categorised into two groups depending on the system architecture: centralized and distributed. In the centralized schemes, a trusted fixed wireless infrastructure, usually a WiFi access point, is employed to check the proximity of mobile users and generate LPs for them. On the other hand, in the decentralized schemes, this task is done by ordinary mobile users who act as witnesses and issue LPs for each other. This makes their implementation easier and cheaper than the centralized mechanisms. In this section, we review the related literature on each category separately.

4.5.1.1 Centralized Schemes

In this approach, a central trusted node such as a wireless access point is utilised to generate LPs for users in a specific site. The idea of employing wireless access points as location proof generators was introduced by Waters et al. [134] for the first time. They measure the round-trip signal propagation latency to decide on the proximity of a user to a trusted access point referred to as the location manager. However, the proposed scheme is vulnerable against relay attacks and specifically against Terrorist Frauds. In other words, their algorithm lacks a mechanism by which the location manager ensures that the received ID is really for the user who has submitted the LP request.

To address this issue, Saroiu et al. [135] proposed a technique in which the access point broadcasts beacon frames consisted of a sequence number. To obtain an LP, users must sign the last transmitted sequence number with their private key and send it back to the access point along with their public key (the access point broadcasts beacons every 100 milliseconds). This makes the system resistant against Terrorist Frauds since the malicious prover does not have enough time to receive the sequence number from the adversary, sign and send it back to the adversary. However, the proposed algorithm has privacy issues because users must reveal their identity publicly. Javali et al. [136] have used the same idea to make their algorithm resistant against relay attacks. They also utilise the unique wireless channel characteristics, i.e., channel state information (CSI) to decide on users' proximity. The proposed scheme consists of three entities, i.e., Access Point, Verifier and Server which makes the system expensive. In addition, the user's identity is revealed publicly which might cause privacy issues. Table 4.4 presents a comparison of these LP schemes.

4.5.1.2 Distributed Schemes

In the distributed scenarios, users collaborate with the system to generate LPs. In other words, users act as witnesses for each other. The main advantage of this approach is that there is no need for a trusted access point to issue LPs. Therefore, this type of systems can be used in locations where users are far from a trusted entity. APPLAUS introduced by Zhu et al. [139] is one of the pioneer research works on distributed location proof systems. In APPLAUS, mobile devices use their short-range Bluetooth interface to communicate with their nearby devices who request an LP. To preserve users' location privacy, they need to select a set of M pseudonyms and change them periodically. These pseudonyms are considered as users' public keys which are required to be registered with a trusted Certificate Authority (CA) along with the associated private keys. However, changing pseudonyms regularly creates a high level of computation and communication overhead. In addition, the users are required to generate dummy LPs as well.

Davis et al. proposed a privacy-preserving alibi (location proof) scheme in [140] which has a distributed architecture. To preserve users' location privacy, in the introduced scheme, their identity is not revealed while an alibi is being created. Thus, only a judge with whom a user submits his/her alibi can see the user's identity. However, collusions and other security threats have not been considered in the work.

In the distributed solutions, Prover-Witness collusions are possible because witness devices are not always trusted. A witness device can issue an LP for a dishonest user while one of them (or both) is not located at the claimed location. This is one of the major challenges of these schemes. For example, in PROPS which has been proposed by Gambs et al. [143], Prover-Witness collusions have not been discussed

Table 4.4 Comparison of LP schemes

LP scheme	Features	Advantages	Disadvantages
Waters et al. [83]	Round-trip signal propagation delay is measured to decide on device proximity	Privacy-aware Lightweight	Vulnerable to P-P collusions
Javali et al. [136]	No DB mechanism is used Utilises channel state information (CSI) to decide on users proximity	Resistant to P-P collusions Fast	Privacy issue Expensive for implementation
Saroiu et al. [135]	Access point broadcasts sequence numbers periodically Provers sign the last transmitted sequence number to request an LP	Resistant to P-P collusions	Privacy issue
VeriPlace [137]	To obtain a final LP, a user needs to get an intermediate LP from a trusted access point	Privacy-aware	Needs three types of trusted entities run by separate parties
STAMP [138]	An entropy-based trust model is used to address P-W collusions	Supports location granularity	Vulnerable to P-P collusions (the broken Bussard-Bagga DB protocol is employed)
APPLAUS [139]	Provers adopt different pseudonyms and change them periodically	Privacy-aware	High communication overheads High computation overheads
Alibi [140]	Provers' ID is revealed only when they choose to submit their alibi to a judge	Privacy-aware Lightweight	Vulnerable to collusion attacks
Link [141]	A group of local users collaboratively verify a prover's location	Resilient to situations when there is not enough neighbour devices	Privacy issue
SPARSE [142]	No DB mechanism is used for secure proximity checking	Resistant to P-P collusions Privacy-aware	Prevents P-W collusions only in crowded scenarios
PROPS [143]	Group signatures and ZKP are used to make provers anonymous	Efficient and privacy-aware	Vulnerable to P-W collusions

although it provides an efficient and privacy-aware platform for users to create LPs for other users.

To the best of our knowledge, there is no efficient and reliable solution proposed in the literature to resolve the Prover-Witness collusions issue with a high level of reliability even though some significant efforts have been made so far. For example, in LINK introduced by Talasila et al. [141] a group of users collaboratively verify a user's location upon his/her request sent through a short-range Bluetooth interface. It is assumed that there is a trusted Location Certification Authority (LCA) to which the verifying users (located in the vicinity of the requesting user) send their verification messages. Then, the LCA checks validity of the claim in case of a Prover-Witness collusion. This is done by checking three parameters: the spatiotemporal correlation between the prover and verifiers, the trust scores of the users, and the history of the trust scores. However, it does not detect and prevent Prover-Witness collusions with a high level of reliability. Moreover, in the LINK scheme, users' location privacy has not been considered in the scheme design since a user needs to broadcast his/her ID to the neighbour verifiers.

STAMP introduced by Wang et al. [138] is another example in which an entropy-based trust model is proposed to address the Prover-Witness collusions issue. This method is also unable to provide the necessary reliability to detect Prover-Witness collusions. In addition, to address Terrorist Frauds, STAMP employs the Bussard-Bagga protocol [144] as the distance bounding protocol which has already been shown to be unsafe [145]. Moreover, the computation time required by STAMP to create an LP is long when users have a large private key [138].

Although different novel methods have been introduced so far, each of them has its own constraints, i.e., privacy issues [135, 136, 141], vulnerability against collusions [134, 138–141, 143], high level of communication and computation overheads [139], and expensive implementation [136, 137]. The scheme proposed in [142] prevents Prover-Witness (P-W) collusions only in crowded scenarios.

4.5.2 Preliminaries

In this section, we first review distance bounding (DB) protocols and present the security attacks that these protocols might experience. These attacks are a threat for location proof systems as well because most LP schemes employ a DB protocol for proximity checking. Following this, we review TREAD and discuss the need of TREAD modification. Furthermore, we present an overview of the blockchain technology and review the three different types of blockchains. Since we introduce a blockchain-based LP scheme in chapter 8, it is needed to review the basic concepts of blockchain systems first. Following this, we present some of the design challenges that we need to address in this part of our research work.

4.5.2.1   Distance-Bounding Protocols

Distance-bounding protocols [144–148], were introduced to determine an upper bound on the distance between a prover and a verifier, whilst at the same time, the prover device authenticates itself to the verifier. In other words, DB protocols aim to provide authenticated proximity proofs in order to prevent some security attacks. Despite some implementation challenges, in the future, DB protocols will be employed by bank payment companies and car manufacturers due to recent advances [145].

All DB protocols work based on the fact that RF signals do not travel faster than light. First, the verifier sends a challenge bit and the prover replies promptly by sending the corresponding response regarding the received challenge bit. This procedure is called *fast bit exchange* in the literature. Then, the verifier measures the related round-trip time (RTT) which must be less than a specified threshold. This threshold is obtained by computing RTT_{max} that is related to the maximum allowed distance to the prover and is obtained through the following equation:

$$RTT_{max} = \frac{2d_{max}}{C} + t_o, \tag{4.70}$$

where d_{max} is the maximum allowed distance, C is the speed of light, and t_o is an overhead time that is added to cover the computation time [145]. This process is repeated n rounds with n different challenge bits (where n is the length of prover's private key). Finally, the verifier either accepts or rejects the prover's claim.

In addition to the proximity checking, the verifier must authenticate the neighbor prover at the same time. Otherwise, an adversary can collude with a remote malicious prover and perform the *fast bit exchange* mechanism on behalf of the remote prover. In this regard, there are some security attacks that a well-designed DB protocol must be resistant against. In the literature, the following security threats have been identified so far [145]. These attacks threat an LP scheme as well since most of the LP schemes employ a DB protocol as their core function.

Distance Frauds: In a distance fraud, a malicious prover tries to convince an honest verifier that his physical distance to the verifier is less than what it really is. This attack can occur if there is no relationship between challenge bits and response bits and the malicious prover knows the time at which the challenge bits are sent. In this case, the malicious prover can send each response bit before its challenge bit is received.

Mafia Frauds: In this attack, an adversary tries to convince an honest verifier that a remote honest prover is in the vicinity of the verifier. The adversary in this attack can be modeled by a malicious prover that communicates with the honest verifier and a malicious verifier who interacts with the honest prover (Fig. 4.26). The car locking system is a good example to understand this type of attacks where an adversary tries to open a car's door by convincing the reader unit that the key is close to the car.

Terrorist Frauds: In this attack, a remote malicious prover colludes with an adversary who is close to an honest verifier to convince the verifier that he/she is in the

Table 4.5 Comparison of the success probability of different security threats for some well-known DB protocols

DB protocol	Distance frauds	Mafia frauds	Terrorist frauds
Swiss-Knife [147]	$(3/4)^n$	$(1/2)^n$ to 1	$(3/4)^{\theta n}$
Gambs et al. [148]	$(3/4)^n$	$(1/2)^n$	1
Bussard-Bagga [144]	1	$(1/2)^n$	1
privDB [150]	$(3/4)^n$	$(1/2)^n$	1
SKI [151]	$(3/4)^n$	$(2/3)^n$	$(5/6)^{\theta n}$
Fischlin–Onete [146]	$(3/4)^n$	$(3/4)^n$	$(3/4)^{\theta n}$

vicinity of the verifier. Although in their collusion, they never share private information (e.g., private key) with each other, it is still possible that they establish a very fast communication tunnel between themselves and the adversary relays the verifier's message to the malicious prover who can sign and send it back to the adversary for submission. Therefore, just a simple assumption that users never share their private key can not protect the system against this type of attacks.

Moreover, there is another attack called *Distance Hijacking* introduced by Cremers et al. [149]. They believe this attack is an extension of distance frauds which is very close to Terrorist Frauds as well. In a distance hijacking attack, a remote malicious prover tries to provide wrong information about his distance to an honest verifier by exploiting the presence of one or multiple honest provers.

To address the mentioned attacks, different DB protocols have been introduced so far [144, 146–148]. However, each protocol has its own constraints (for more detail refer to [145]). For example, the popular Bussard-Bagga protocol (introduced by Bussard et al. [144] to address the Terrorist Frauds) was proven insecure by Bay et al. [145]. Table 4.5 compares some well-known DB protocols in terms of vulnerability against the mentioned security threats and frauds. In this table, the success probability of the most common security threats have been shown. n indicates the number of rounds in the DB process and θ is a parameter related to a Terrorist Fraud such that it is difficult to prevent from the exhaustive searches that are done to recover θn bits (see [145] for more details).

As we see in the table, most of the DB protocols are vulnerable to at least one security attack. Moreover, the two fraud-resistant protocols, i.e. SKI [151] and Fischlin–Onete [146], need a large n to provide sufficient reliability which makes the DB process slow since the process is performed for n rounds.

4.5.2.2 TREAD

TREAD is a secure and light-weight DB protocol proposed by Avoine et al. [152] to address the aforementioned problems. In TREAD, a novel idea has been deployed to make the protocol resistant to Terrorist Frauds: if a dishonest prover colludes

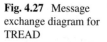

Fig. 4.27 Message exchange diagram for TREAD

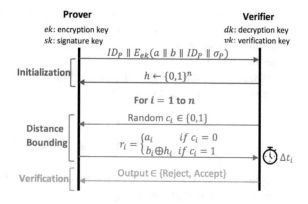

with another user to conduct a Terrorist Fraud, he can be easily and unlimitedly impersonated by the accomplice later. This risk is not easily taken by any rational prover.

Assuming there is a prover device in the vicinity of a trusted verifier who have secretly shared the encryption/decryption key pair ek and dk, and the signature/verification key pair sk and vk, TREAD is performed in three phases, i.e. *Initialization*, *Distance Bounding*, and *Verification* (Fig. 4.27).

(1) *Initialization:* In this phase, the following activities are performed by the prover and verifier devices:

Prover: The prover device generates two random bit-strings a and b from the uniform distribution on $\{0, 1\}^n$, computes the signature $\sigma_P = S_{sk}(a||b||ID_P)$ and the encrypted message $e = E_{ek}(a||b||ID_P||\sigma_P)$ where ID_P is the prover's ID (see Table 4.6 for a list of notations). Then, it sends $e||ID_P$ to the verifier.

Verifier: Upon receiving $e||ID_P$, the verifier device decrypts e using the decryption key dk and checks the prover's signature σ_P using the verification key vk to see if it is correct. If σ_P matches the prover's signature, the verifier generates a random bit-string h from the uniform distribution on $\{0, 1\}^n$ and sends it to the prover.

Table 4.6 List of cryptography notations

Notation	Description
$\|$	Concatenation
$S_{ent}(m)$	Signature of entity *ent* on message m
$E_{ent}(m)$	Encryption of message m using public key of entity *ent*
Loc	GPS coordinates related to the prover's Location
ID_P	The prover's identity
ID_W	The witness's identity
$\oplus$	XOR operation

(2) *Distance Bounding:* In this phase, the prover and verifier devices start to perform the n-stage fast bit exchange process :

Verifier: In stage i, $(i = 1, 2, \ldots, n)$, the verifier picks a random bit c_i, sends it to the prover and starts its timer.

Prover: Upon receiving c_i, the prover immediately computes the following bit $r_i =$ and sends it back to the verifier:

$$
r_i = \begin{cases} a_i, & \text{if } c_i = 0 \\ b_i \oplus h_i, & \text{if } c_i = 1 \end{cases} \tag{4.71}
$$

Verifier: When r_i is received by the verifier device, it stops the timer and records its value Δt_i. Then, it performs stage $i + 1$ until all the n stages are done after which it goes to the verification phase.

(3) *Verification:* In this phase, the verifier device checks all the received r_i for $i = 1, 2, \ldots, n$ to see if they have been correctly computed based on h_i, c_i, a_i and b_i (the last two bits received in the initialization phase). Then Δt_i must be less than the predefined threshold RTT_{max} for $i = 1, 2, \ldots, n$.

Finally, the prover's request is accepted if the above checkings are successfully passed for all n stages.

As we see, in case of a Terrorist Fraud, a dishonest prover (located far from the verifier) not only has to provide the accomplice with his σ_P and e, but also his random bit-strings a and b. Otherwise, the accomplice is unable to correctly respond to the challenge bits c_i in the DB phase. This enables the accomplice to easily impersonate him later using a, b, σ_P, and e. See [152] for a comprehensive security analysis on TREAD.

4.5.2.3 TREAD Modification

In spite of the security guarantees that TREAD offers, it needs some amendment before we make use of it in our proposed architecture. In the following, we show how the prover's location privacy is negatively affected, if TREAD is integrated into PASPORT without any customization.

In TREAD, the prover's ID is sent to a neighbor verifier (which is assumed to be trusted) through a short-range communication interface. Due to PASPORT's decentralized architecture, the trusted verifier is located far away from the prover. Instead, a witness device (which is untrusted from a privacy point of view) collects the prover's data and performs the DB procedure. Thus, the prover's ID is sent to the witness devices in the form of a plain text message if we integrate TREAD into PASPORT without any modification. This breaches the prover's location privacy. Hence, it is necessary to modify TREAD and make it a privacy-aware DB protocol.

Note that prover anonymity can be offered by TREAD if group signatures are used [152]. However, they guarantee provers' anonymity up to group level only. Since we do not want to use group signatures in the PASPORT's architecture, in the next section, we propose a private version of TREAD, i.e. P-TREAD, by which a prover device can anonymously broadcast its LP request for neighbor witnesses while he/she benefits from the TREAD security guarantees.

4.5.3   PASPORT: The Proposed Scheme

In this section, we present Privacy-Aware and Secure Proof Of pRoximiTy (PAS-PORT), which performs LP generation and verification for mobile users in a secure and privacy-aware manner. The proposed scheme provides the integrity and non-transferability of generated LPs. To make PASPORT resistant to P-P collusions and perform private proximity checking, we develop a privacy-aware distance bounding (DB) protocol P-TREAD and integrate it into PASPORT. P-TREAD is a modified version of TREAD [152], a state of the art and secure distance bounding protocol without privacy consideration. Our customization does not affect TREAD's main structure and features. Thus, PASPORT benefits from its security guarantees [50]. By employing P-TREAD as the distance bounding mechanism, a malicious prover colluding with an adversary can easily be impersonated by the adversary later. Generally, users do not take such a risk by initiating a Prover-Prover collusion. In addition, to resolve the P-W collusions issue, we propose a witness selection mechanism that randomly assigns the available witnesses to the requesting provers instead of allowing them to choose the witnesses themselves. We show that by adopting this mechanism, a P-W collusion can be conducted with only a negligible success probability if LBSPs create sufficient incentives for users to act as witnesses and generate LPs for provers.

In this section, we *firstly* present the PASPORT framework and its entities. *Secondly*, we present the trust and threat model which we have considered in our work. *Following this*, we introduce P-TREAD. *Finally*, the full framework of the PASPORT scheme is presented.

4.5.3.1   Architecture and Entities

The proposed system architecture is shown in Fig. 4.28. As we see, the system has a distributed architecture and consists of three types of entities, i.e., Prover, Witness and Verifier. A *Prover* is a mobile user who requires to prove his/her location to a verifier. A *Witness* is the entity that accepts to issue an LP for a neighboring prover upon request. We assume service providers create sufficient incentives for mobile users to become a witness and certify other users' location. In PASPORT, we consider witnesses as mobile users.

Finally, a *Verifier* is the unit who is authorized by the service provider to verify LPs claimed by provers. We assume provers communicate with witnesses through a short-range communication interface such as Wi-Fi or Bluetooth. This short-range communication channel is supposed to be anonymous such that users can broadcast

Fig. 4.28 The proposed
system architecture

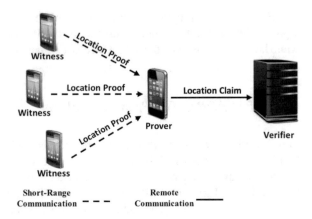

their messages over it without revealing their identifying data such as IP or MAC
address.

4.5.3.2   Trust and Threat Model

We assume mobile users are registered with the service provider. Each user has a
unique public-private pair key stored on his/her mobile device and certified by a
Certification Authority (CA). Users' identity is determined through their public key
and we assume users never share their private key with other users because they do
not give their mobile devices to others [136, 138]. Thus, in a collusion scenario, we
suppose a malicious prover never goes that far to provide another party with his/her
private key. We also assume all the messages exchanged between the entities might
be eavesdropped by passive eavesdroppers. In the following, we discuss the trust and
threat model for each entity individually.

Prover. It is assumed that the prover makes an effort to obtain false LPs. This can
be done through different scenarios in which a prover might (a) try to provide the
witnesses with fake information about his/her location to convince them to generate
LPs for him/her, (b) manipulate the LP issued for him/her to change its location or
time field, (c) attempt to steal an LP issued for another user and use it for him/herself,
and (d) collude with other users (provers or witnesses) to obtain LPs. Moreover, we
assume provers try to obtain the identity of witnesses.

Witness. A witness might collude with a prover to generate a fake LP for him/her.
In addition, a witness may try to deny an LP which has been issued by him/herself.
Witnesses are assumed to be curious about the provers' identity.

Verifier. We suppose the verifier is trusted and never leaks users' identity and
their spatiotemporal data. It is assumed that the verifier keeps a regularly updated
list of witnesses who are present at the given location and have accepted to generate
LPs for other users. The verifier accepts the LPs issued by these witnesses only. We
suppose service providers create necessary incentives to encourage selfish users to

collaborate with the system. Otherwise they might not generate LPs to save their battery power or reduce their communication costs.

Regarding collusions, we consider both Prover-Prover and Prover-Witness collusions in our threat model as it can be directly derived from the above assumptions. In the next subsection, we introduce the proposed privacy-aware DB protocol P-TREAD.

4.5.3.3 P-TREAD

In this subsection, we present P-TREAD, a modified version of TREAD, for private proximity checking in the PASPORT architecture.

As discussed in the Preliminaries subsection, to protect users' privacy, we need to customize TREAD in such a way that provers can anonymously submit an LP request to neighbor witnesses. For this reason, in P-TREAD, we limit a witness' role to only collecting (not verifying) the required data from the prover (the verification is performed by the remote trusted verifier). All the privacy-sensitive data are encrypted by the prover and sent to a witness who signs and sends them back to the prover as an LP. Then, after the claim (received LP) is submitted to the verifier by the prover, verification of the claim can be performed by the trusted verifier in the next phase. We divide the whole procedure into two phases, (a) data collection and LP generation, and (b) authentication and verification.

Phase 1. Data collection & LP generation. In this part of the protocol, the initialization phase of TREAD is performed with the following exceptions:

- The prover device does not send ID_P to the witnesses as a plain text message (it only sends e to the witnesses).
- e is computed by the prover device using the verifier's public key. Therefore, the witnesses can not decrypt it and deanonymize the prover. We assume that the verifier publishes its public key for the users. Moreover, every user has registered a public/private key pair with the verifier.
- The witness devices do not check the prover's signature σ_P since the prover must be anonymous (in addition, they can not decrypt e and obtain the signature). Later, σ_P will be checked by the verifier in the next phase.

Then, the DB procedure is performed similar to the DB phase of TREAD by which the prover's responses r_i to challenge bits c_i ($i = 1, 2, \ldots, n$) are collected. After data collection is finished, the witness device creates the following LP and sends it to the prover:

$$LP = E_{Verifier}(m_2 || S_{Witness}(m_2)) \tag{4.72}$$

where $m_2 = r || c || h || e || ID_W || Loc || time$ and ID_W is the witness ID. Note that the prover can not see ID_W since it is encrypted using the verifier's public key. This preserves location privacy of the witnesses as well. Finally, the prover submits the following message with the remote verifier:

$$LP' = E_{Verifier}(LP||a||b||ID_P) \tag{4.73}$$

In other words, the witness collects the required information from the prover (e and r), creates message LP, and sends it to the prover for submission. In this phase, the witness can not see the prover's identity as it has been encrypted by the verifier's public key in message e.

Phase 2. Authentication & verification. In this phase, the verifier authenticates the prover based on the received LP' and verifies the validity of the LP issued by the witness. To do this, it first decrypts LP' using its private key and extracts LP, a, b, and ID_P. Then, it checks the following:

- The signature σ_P placed in message e must match the prover's signature based on ID_P.
- The received ID_Ps placed in e and LP' must match.
- The witness signature on message m_2 must match the signature associated with ID_W.
- The two $a||b$s placed in the messages e and LP' must match.
- The received response r must match r' where r' is obtained based on the received a, b, h, and c bit-strings.

If all the above checks are successfully passed, the prover's location claim is accepted by the verifier.

As we see, by using P-TREAD, a prover can anonymously request an LP from neighbor witnesses while the main structure of TREAD is preserved which brings security guarantees for users. In the next subsection, we integrate P-TREAD into our main LP scheme, i.e. PASPORT, to perform secure and private device proximity checking.

4.5.3.4   The Workflow of PASPORT Framework

The proposed LP scheme consists of three main phases: *Initialization*, *Location Proof Generation*, and *Location Claim and Verification* (Fig. 4.29).

(1) Initialization: In this phase users register with the system and the Certification Authority certifies users' public-private key pairs. Moreover, the verifier creates a Witness Table in which it keeps the identity and location of mobile users who accept to be a witness. This table is regularly updated as witnesses sign on or off at every site. Furthermore, for every registered user in the system, the verifier records a list of provers for which the user generates an LP. These lists are used by the verifier to select which witnesses are qualified to generate LPs for a specific prover. This is done to prevent Prover-Witness collusions.

(2) Location Proof Generation: This phase is run in two stages: *Witness Selection* and *P-TREAD Execution*.

(2.1) Witness Selection: In this stage, the prover submits an LP request to the verifier. Upon receiving the prover's request, the verifier selects K witnesses form

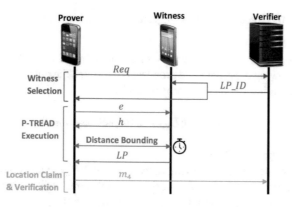

$Req = E_{Verifier}(ID_P \parallel Loc)$

$e = LP\_ID \parallel E_{Verifier}(m_1 \parallel S_{Prover}(m_1))$, $m_1 = a \parallel b \parallel ID_P \parallel Loc$

$LP = E_{Verifier}(m_2 \parallel S_{Witness}(m_2))$, $m_2 = r \parallel c \parallel h \parallel e \parallel ID_W \parallel Loc' \parallel time$

$m_4 = E_{Verifier}(LP_1 \parallel LP_2 \parallel \cdots \parallel LP_K \parallel a \parallel b)$

Fig. 4.29 Message flow between the three entities of the proposed scheme

its Witness Table to generate LPs for the prover. This is done to neutralize Prover-Witness collusions because in this case, the prover does not have control over the witness selection process. However, to further protect PASPORT against prover-witness collusions, we integrate an entropy-based trust model as a supplementary method into the witness selection mechanism. Using this trust model, a trust score is computed by the verifier for every available witness device w based on its LP generation history and the number of LPs that w and the prover have issued for each other in the past. If the obtained score is above a threshold, the device is selected to witness for a requesting prover. The following step by step activities are performed in this stage:

1. **Prover**: First, the prover sends the following message Req to the verifier to inform it that he/she wants to start requesting an LP. This message can be sent to the verifier through the prover's Internet connection.

$$Req = E_{Verifier}(ID_P \parallel Loc) \tag{4.74}$$

2. **Verifier**: Upon receiving the prover's message, the verifier extracts all the witnesses who have recently (in a reasonable period of time) proved that they are in an acceptable distance to location Loc from its Witness Table (this acceptable distance is defined depending on the application). Then, K witnesses are selected among the shortlisted witnesses using the proposed trust model. These K witnesses are then qualified to generate LPs for this prover. If there are not enough qualified witnesses, the verifier suspend this request until the necessary number of qualified witnesses become available. Then, the verifier generates a unique ID for this LP (LP_ID) and sends it to the selected witnesses and the prover as well.

(2.2) P-TREAD Execution: In this stage, the prover starts to perform the P-TREAD protocol.

1. **Prover**: The prover generates two n-bit random numbers a and b, and then computes the following message e and broadcasts it through the predefined short-range communication interface (WiFi or Bluetooth).

$$e = LP\_ID \| E_{Verifier}(m_1 \| S_{Prover}(m_1)),\qquad(4.75)$$

   where $m_1 = a \| b \| ID_P \| Loc$.
2. **Witness**: A witness upon receiving e, extracts the LP_ID and compares it with the one received from the verifier. If they are not same, it discards e. Otherwise, it generates an n-bit random number h and sends it to the prover.
3. **Prover**: The prover computes $(z_i = b_i \oplus h_i)$ for $i = 1, 2, \ldots, n$ and sends an Ack to the witness.
4. **Witness**: The witness starts an n-stage time sensitive DB process by generating a random bit c_i at each stage i and sending it to the prover. It also starts a timer immediately after sending c_i.
5. **Prover**: Upon receiving c_i, the prover instantly sends the following response r_i to the witness:

$$r_i = a_i.\bar{c}_i + z_i.c_i\qquad(4.76)$$

6. **Witness**: The witness stops the timer when the response r_i is received from the prover. The timer must show a time less than the predefined threshold $\frac{2d_{max}}{C} + t_o$, where d_{max} is the maximum allowable distance between the prover and the witness, C is the speed of light, and t_o is the overhead time required by the prover to compute the response bit r_i upon receiving c_i. If all the n responses are received in the correct time, the witness issues the following location proof and sends it to the prover:

$$LP = E_{Verifier}(m_2 \| S_{Witness}(m_2)),\qquad(4.77)$$

   where $m_2 = r \| c \| h \| e \| ID_W \| Loc' \| time$.
   For timer values larger than this threshold, the witness generates the following location proof:

$$LP = E_{Verifier}(m_3 \| S_{Witness}(m_3)),\qquad(4.78)$$

   where $m_3 = ID_W \| reject$.

As we see, we adopt the *sign-then-encrypt* model to compute PASPORT messages. This protects the privacy of provers (witnesses). The reason is that if the more common *encrypt-then-sign* model is chosen, a witness (prover) can check the signature on e (on LP) with the public keys of all the users and find the prover's (witness') identity. Moreover, by using this method, eavesdroppers never infer the users' identity.

(3) Location Proof Claim and Verification: Upon receiving LPs from all the K witnesses, the prover concatenates them in message m_4 and sends it to the verifier.

$$m_4 = E_{Verifier}(LP_1\|LP_2\| \ldots \|LP_K\|a\|b) \tag{4.79}$$

The verifier checks the received location proofs and either accepts or rejects the prover's claim. First, it decrypts each witness's location proof and message e using its private key. Then, it computes $r_i' = a_i.\bar{c}_i + (b_i \oplus h_i).c_i$ for $i = 1, 2, 3, \ldots, n$ regarding the received a, b, c, and h. If $r_i' \neq r_i$, the verifier rejects the prover's claim. Otherwise, the following checks are performed by the verifier:

- Is the witness with identity ID_W among the witnesses which have been qualified by the verifier in the Witness Selection stage?
- Are the two ID_Ps extracted form Req and m_1 the same?
- Are prover's and witnesses' signatures on m_1 and m_2 correct regarding ID_P and ID_Ws respectively?
- Is Loc in an acceptable range of Loc'?
- Is $time$ in an acceptable range of the current time?
- Are the two $a\|b$ s received in the messages m_1 and m_4 the same?
- Is $K - K_R \geq T$ correct? Where K_R is the number of rejected location proofs and T is a threshold which is defined depending on the application.

Assuming the prover's location claim passes all the above checks successfully, the verifier accepts the prover's claim.

4.5.3.5 Witness Trust Model

To further protect PASPORT against prover-witness collusions, we integrate an entropy-based trust model into the PASPORT witness selection mechanism. Using this trust model, the verifier computes a trust score for a witness device based on its LP generation history. If the obtained score is above a threshold, the device is selected to witness for a requesting prover. In fact, a witness device receives a low score if it has issued many LPs for that prover. Thus, the prover device is prevented from receiving its LPs from a small group of witnesses only.

We adopt an entropy-based approach to measure the trust scores. In information theory, entropy represents the average amount of information that we get from a message produced by a stochastic source of data. It works based on the fact that when a low-probability message is received, it carries more information than when the source of data produces a high-probability message. Thus, it is a suitable measure of the level of diversity and randomness that a prover device should have in the list of its witnesses. In other words, when a higher entropy is obtained for a witness device, it is more likely that it has generated LPs for a diverse range of provers rather than for a small group of prover devices.

Consider w is a witness device that has already issued at least one LP for N prover devices $p_1, p_2, \ldots, p_N$. Assume $A(w, p_i)$ is a percentage of the past LP transactions between w and p_i out of w's total past LP transactions. The entropy of w is obtained using the following equation.

$$e_w = -\sum_{i=1}^{N} A(w, p_i) log(A(w, p_i)) \tag{4.80}$$

We define $S(w, p_i)$ as the trust score of device w to be selected as a witness for the prover device p_i.

$$S(w, p_i) = \frac{e_w e_{p_i}}{1 + B(w, p_i)}, \tag{4.81}$$

where $B(w, p_i)$ is the number of LPs that w and p_i have issued for each other in the past out of their total number of LP transactions.

As a result, in the witness selection phase, the verifier selects those devices with the highest trust score. This prevents the system from selecting a witness who may have a connection with the prover and has already issued several LPs for the prover. Moreover, using this model, possible prover-witness collusions can be detected and prevented by the system because a prover device who has received majority of its LPs from a small group of witnesses is more likely to collude with these witnesses.

4.5.3.6 PASPORT Usability

Since PASPORT has a decentralized architecture, it relies on the collaboration of mobile users to generate location proofs for each other. Note that users usually need to have an LP for crowded public places (e.g., shopping centers). This mitigates the concerns about the number of available witnesses. However, mobile users may refuse to collaborate with the system in order to save their battery power or reduce their communication costs. To address this issue, service providers should create sufficient incentives for mobile users to collaborate with the system and certify other users' location. In this regard, we propose two approaches to overcome the issue.

1. Location-based service providers can incentivize mobile users to collaborate by offering them some rewards, badges and benefits that they are currently providing to their users. These rewards can be granted to mobile devices based on their contribution in the network, e.g., the number of LPs that they have generated for other users in a given time period. Moreover, other businesses such as insurance companies and government agencies that might utilize LPs of their customers can contribute to make the rewards more valuable. This creates the necessary incentive for mobile users to collaborate with the system.
2. The second approach is to integrate an incentive mechanism into the proposed scheme, e.g., using a blockchain architecture that remunerate users with a given amount of a cryptocurrency. Since PASPORT is a decentralized scheme, the distributed architecture of blockchains is an appropriate platform to address this issue. This encourages mobile users to collaborate with the system and respond to other users' LP requests.

Thus, by applying the incentive policies on the proposed scheme and encouraging mobile users, a sufficient number of witness devices are become available for the verifier to select.

4.5.4 Security Analysis

In this section, we perform a comprehensive security and privacy analysis to show that PASPORT achieves the necessary security and privacy properties of a secure and privacy-aware LP scheme described in [153].

1. Resistance to Distance Frauds: In PASPORT, distance frauds are prevented by the time sensitive DB process (performed in stages 2–2–d, 2–2–e and 2–2–f) which is performed via a short-range communication interface. Moreover, the existence of the random number h ensures us that $a_i \neq b_i \oplus h_i$. Otherwise (if the witness does not send h and the prover responds just with $r_i = a_i.\bar{c}_i + b_i.c_i$), a remote malicious prover can simply select $a = b$ and send $r_i = a_i = b_i$ before it receives the challenge bit c_i (in this case r_i does not depend on the challenge bit c_i). Thus, for a remote malicious prover the only way to have his fake location verified is colluding with a dishonest prover or witness. As we see in this section, PASPORT is resistant to Prover-Prover collusions and reduces the success probability of Prover-Witness collusions to a negligible value.

2. Unforgeability: It is not feasible for a malicious prover to create a location proof himself without proving his location to a qualified witness. The reason is that the verifier checks each qualified witness' ID with their signature on m_2. Since users do not share their private key with each other, the malicious prover can not create the witness' signature on m_2 even if he knows the identity of each qualified witness. Moreover, an adversary who tries to forge another user's location proof will not be successful because he does not have the victim's private key to sign m_1. Furthermore, if a location proof is created by a dishonest witness W' who has not been selected as a qualified witness, it will be easily detected by the verifier by comparing the identity of W' with all the qualified witnesses' identity.

3. Non-Transferability: Suppose an adversary wants to use a location proof which has been issued for prover P. Even if the adversary knows the prover's ID, i.e., ID_P, he still does not have the random numbers a and b to create m_4 and submit his claim. Note that random numbers a and b have been encrypted using the verifier's public key. Thus, neither the adversary nor the witness can see them. Moreover, the presence of $time$ in m_2 makes it infeasible for the prover device P to give its location proof along with ID_P, a and b to another device for later submissions. In this case, the prover can not change $time$ because it has been signed by the witness' private key and encrypted using the verifier's public key.

4. Resistance to Mafia Frauds: Suppose an adversary A wants to perform a Mafia Fraud on prover P and witness W who are both honest. Suppose A consists of (or is modeled by) a witness $\bar{W}$ and a prover $\bar{P}$. Even if we assume that $\bar{W}$ obtains message e by communicating with P, it is not feasible for $\bar{p}$ to fool W using e. The reason is that $\bar{P}$ must successfully perform the DB process by sending response bits r_i to W. This requires the total knowledge of random numbers a and b which P never sends them to $\bar{W}$. Moreover, A does not gain any further knowledge about a and b by pretending to be different witnesses (for example n malicious witness $\bar{W}_1, \bar{W}_2, ..., \bar{W}_n$). This is because P generates different numbers a and b whenever he/she performs P-TREAD. In other words, for different witnesses, different $a\|b$ is generated. Therefore, PASPORT is resistant to Mafia Frauds.

5. Resistance to Terrorist Frauds (Prover-Prover collusions): Suppose a remote malicious prover P colludes with an adversary A which is close to an honest witness W to obtain a fake LP. In this attack, A must send message e to W and perform DB process on behalf of P. To perform this attack, P helps A by generating message e and sending it to A. In addition, P has to send the random numbers a and b to A as well. Otherwise, A can not respond to challenge bits c_i in DB process and the attack is defeated. However, if P sends $a\|b$ to A, he can easily impersonate P later for as many times as he wants. Therefore, the prover must select one between performing the attack and being impersonated. In fact, in PASPORT, the cost of a Prover-Prover collusion is increased to such a level that no rational prover accepts its risk.

6. Resistance to Sybil Attacks: In a Sybil attack, an adversary tries to control or influence a peer-to-peer network by creating multiple fake identities. There are a number of countermeasures that can be adopted to make PASPORT resistant to Sybil attacks.

1. **Identity Verification:** Since PASPORT is a permissioned peer-to-peer network (rather than a permissionless network, e.g., Bitcoin), all users' identities are verified before they are authorized to access the system. This can be supported by forcing users to perform a two-factor authentication process when they register to the network. For example, users may be asked to provide a security code sent to their mobile phone or email address. In this case, the network rejects to create a new account if a duplicate mobile phone number or email address is provided by the adversary. This makes a Sybil attack non-economic for malicious users since they have to provide many SIM cards or email addresses to proceed with the attack. Alternatively, users may need to sign up using individual email addresses or social network profiles, e.g., Facebook accounts. Furthermore, in a specific time interval, no more than a specific number of accounts may be allowed to be created using a single IP address.

2. **Unequal Reputation:** A supplementary technique to prevent Sybil attacks is to consider different levels of reputation for different accounts. Using this technique, witness devices associated with the accounts with an older creation date receive more reputation and their testimony is highly accepted. Newly-created accounts must remain active for a specific period before they become eligible to witness. This limits the power of new accounts. Therefore, creating many new accounts does not result in any advantage for a Sybil attacker against other older reputable accounts.

3. **Cost to create an identity:** To prevent malicious users from creating multiple fake accounts, the network may consider a small cost for every user that wants to join the network. In this case, the cost to create a single account is small. However, the total cost to create many identities is higher than the reward or benefit that the attacker receives after successfully conducting the Sybil attack. Note that it is more important to make it expensive for an attacker to create and control multiple accounts in a short period of time rather than just creating a new account. In other words, considering a cost for identity creation should not restrict honest users from joining the network. In fact, the amount of cost should be selected in such a way that creating many accounts becomes non-economic comparing to the benefits that the attacker receives.

7. Resistance to Witnesses Collusions: Witnesses might collude to obtain an honest prover's e and $a\|b$ to impersonate P later. Since P generates different random numbers a and b each time he/she communicates with a witness, the colluding witnesses do not gain more information than what they could obtain without collusion.

8. Preventing Prover-Witness Collusions: In PASPORT, using the witness selection mechanism, the verifier qualifies some witnesses to generate LPs for a prover. A list of these qualified witnesses is kept and linked to the LP_ID by the verifier. Later, in the claim verification phase, the verifier rejects those location proofs generated by unqualified witnesses. Therefore, a malicious prover can not select a specific witness to generate an LP for him.

Let's consider a case in which a remote malicious prover P colludes with some dishonest witnesses which are present at the desired location. We assume K_D is the number of these colluding witnesses who have not generated an LP for P before in a specific period of time. Now, suppose $N > K_D$ is the total number of witnesses who have accepted to collaborate with the system at this location (including the dishonest witnesses) and have not generated an LP for P since a specific time. Note that creating necessary incentives for the witnesses by the service provider can make N a large number. In PASPORT, a location claim is accepted if there are at least T valid (non-rejected) location proofs associated with the claim. Thus, for $K_D \leq T$ the attack is definitely defeated. If $T \leq K_D \leq K$ and x is the number of dishonest witnesses who have been qualified and selected by the verifier, the success probability of a Prover-Witness collusion is obtained through the following equation:

$$
\begin{aligned}
P_{success} &= P(x \geq T) \\
&= P(x = T) + P(x = T + 1) + \ldots + P(x = K_D) \\
&= \sum_{j=T}^{K_D} P(x = j) = \frac{\sum_{j=T}^{K_D} \binom{K_D}{j}\binom{N-K_D}{K-j}}{\binom{N}{K}}
\end{aligned} \tag{4.82}
$$

Note that the malicious prover has to collude with the witnesses who are physically present at the location. This makes it very difficult to have a large K_D. However, we assume he can select $K_D \geq K$. In this case, if $T = K$ is selected by the system, we have:

$$P_{success} = P(x = T) = \frac{\binom{K_D}{K}}{\binom{N}{K}} = \frac{K_D!(N - K)!}{N!(K_D - K)!} \qquad (4.83)$$

Figure 4.30 shows the collusion success probability for different K_D and system parameters. As we see, if $K \geq 0.5N$ is selected, the success probability of a collusion is always less than 0.03. In STAMP [138], a similar LP scheme, the system will detect collusions with a 0.9 success rate if a malicious prover P colludes with 5% of all the users. Note that in STAMP, P can select any user to collude with, no matter where he/she is located. In PASPORT, if P colludes with approximately 50% of the witnesses who are physically present at the desired location and have not generated an LP for him before, the system can prevent this collusion with a success rate better than 0.97. Obviously, the second situation which offers a better prevention rate is much tougher for P to fulfill. Therefore, with carefully chosen parameters, PASPORT provides a more reliable solution for Prover-Witness collusions than what is proposed in STAMP.

9. Resistance to Distance Hijacking: In distance hijacking attacks, a remote malicious prover H tries to fool an honest witness W on their mutual distance by using the involuntary help of an honest prover P which is close to W. Suppose H initiates the protocol by sending Req to the verifier. Upon receiving the related LP_ID, H must broadcast his message e_H through a short-range interface but he is not physically close enough to the qualified witnesses to do so. Thus, the attack can not proceed. Even if we assume that H broadcasts e_H for the witnesses, the attack is defeated. The reason is that in this attack it is assumed that P responds to W's challenge bits in the DB process since H is remote. However, the honest prover P is not aware of random numbers a_H and b_H by which H has already created e_H. Instead, P replies to W with his/her own response bits r_i computed using P's random numbers a_P and b_P in the message e_P. This causes W to generate the LP based on e_P other than e_H. Therefore, if H uses the generated LP to submit his claim with the verifier, this claim will be rejected since the signature on m_1 (in e_P) does not match with H's identity. If H sends his $a_H \| b_H$ to P beforehand, the Distance Hijacking attack converts to a Terrorist Fraud in which P colludes with the remote malicious prover. As we discussed before, PASPORT is resistant to Terrorist Frauds as well.

10. Prover Location Privacy: The prover's ID appears in messages Req, m_1 and m_4. These messages are encrypted by the verifier's public key. Thus, the verifier is the only entity who can identify the prover and neither the witnesses nor an eavesdropper can see the prover's ID. As we discussed before, the *sign-then-encrypt* model improves PASPORT's ability to preserve user's location privacy.

11. Witness Location Privacy: Since a witness device encrypts its ID using the verifier's public key, it is not feasible for the prover or an eavesdropper to identify the witness. Also, users' signatures do not reveal their identity because of the employed *sign-then-encrypt* model.

Fig. 4.30 Success probability of a Prover-Witness collusion for different values of K_D and system parameters

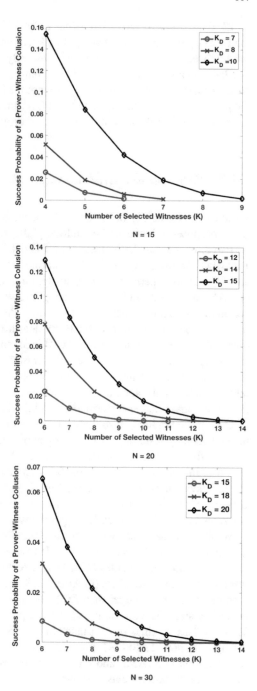

12. Resistance to Eavesdropping: In PASPORT, the prover and witness encrypt their messages with the verifier's public key. Therefore, an eavesdropper gains nothing by listening to their communications. Only LP_ID is sent without encryption that has no value by itself. Moreover, obtaining message e without the total knowledge of random numbers a and b does not enable an eavesdropper to impersonate the prover later. In addition, since PASPORT provides non-transferability, an eavesdropper can not make a claim with an eavesdropped LP issued for another user.

4.5.5 Performance Evaluation

To study the feasibility of the proposed scheme, we implemented a Java prototype of the proposed scheme on the Android platform. Our experiments were performed on two Android mobile devices: (1) a LG G4–H818P equipped with a Hexa-Core 1.8 GHz processor, 3 GB of RAM, and running Android OS 5.1, acting as a prover, and (2) a Sony Xperia Z1 equipped with a Quad-Core 2.2 GHz processor, 2 GB of RAM, with Android OS 4.4.4, acting as a witness. We adopted Bluetooth as the communication interface between the mobile devices and conducted the tests in both indoor and outdoor environments. Each measurement shown in this section has been obtained by averaging the results of 10 independent tests. We used RSA key pairs for encryption and SHA1 as the one-way hash function to compute users' signatures. Since the LP verification phase is performed by the verifier server that has a high level of storage and computational power, we focus our experiments on the P-TREAD Execution phase that is performed by mobile devices with limited resources.

During the application runtime, we measured the CPU utilization of the implemented code by installing a monitoring application that reports the amount of CPU usage of the processes running on the device. As we see in Fig. 4.31a, the CPU usage for a user in standby mode is almost 0.5% and independent of the key size. However, due to heavy computations required for encryption and signature calculations in the LP generation phases, the average CPU usage increases to 2.5%, 8%, and 19% for key sizes 1024, 2048, and 3072, respectively.

We also recorded the amount of time that PASPORT requires to generate an LP after the prover device receives LP_ID from the verifier. We compared the results to the decentralized schemes STAMP and APPLAUS. Figure 4.31b and c show the results for different key sizes (in APPLAUS, the authors have not implemented their scheme for key sizes larger than 256). As expected, longer times were recorded for larger key sizes. The reason is that the DB phase is performed for n challenge bits. Thus, for larger values of n, it takes longer for the DB phase to be performed. As the figures show, PASPORT provides faster responses than similar schemes. The reason is that in STAMP and APPLAUS, the Bussard-Bagga DB protocol is used for provers' proximity checking while in PASPORT, we integrate P-TREAD into the scheme to perform this job that is a more lightweight protocol regardless of its security advantages over the Bussard-Bagga protocol. Unlike P-TREAD, in the Bussard-Bagga protocol, different commitment and decommitment computations are

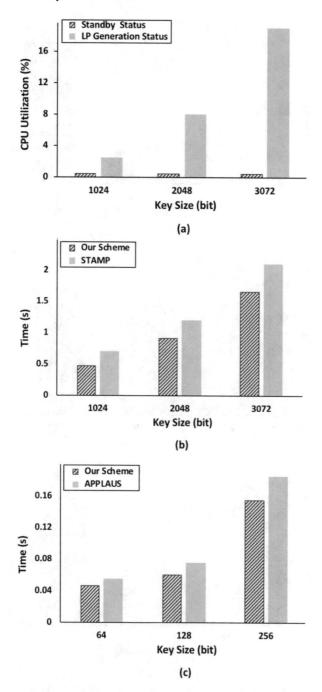

Fig. 4.31 **a** CPU usage for different key sizes. **b** and **c** Time required for LP generation in our scheme, STAMP [138], and APPLAUS [139] under different key sizes. In APPLAUS, the authors have not implemented their scheme for key sizes larger than 256

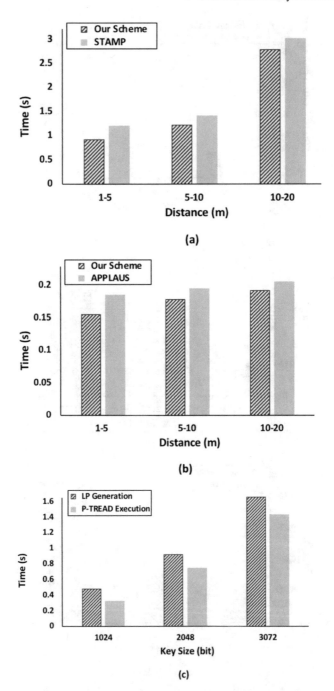

Fig. 4.32 a and **b** Time required for LP generation over different physical distances. The shown measurements are for the key sizes 2048 for (**a**) and 256 for (**b**). **c** P-TREAD distance bounding protocol takes most of the time required for LP generation

needed to be performed by the prover and witness devices, respectively. Moreover, STAMP requires to perform at least two commitment calculations in order to provide location privacy. In APPLAUS, to preserve users' location privacy, they need to select a set of M pseudonyms and change them periodically. This creates a high level of computation and communication overhead.

To evaluate the impact of physical distance between the mobile users on LP generation, we conduct our experiments for different distances and compare the results to the performance of STAMP and APPLAUS. As we see, for longer distances, the required time for PASPORT to generate an LP increases since higher communication latencies occurring in this case. Note that distance only affects the Bluetooth communication latency and does not change the amount of time required for computations performed in mobile devices.

Finally, Fig. 4.32c shows what percentage of the time required for LP generation is taken by the P-TREAD Execution phase. As we see, most of this time is taken by the DB protocol since it requires multiple Bluetooth transmissions. As we discussed before, this time is increased for larger key sizes. As a result, the selection of key size has a critical impact on the scheme's performance. Although larger key sizes provide stronger security, they impose more computational and storage overheads.

Fig. 4.33 Outdoor path for the mobility tests (300 m)

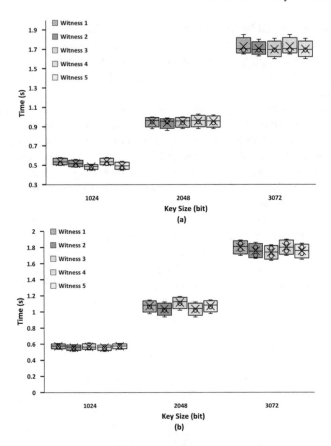

Fig. 4.34 Time required for LP generation when multiple witness devices are involved. **a** Outdoor and **b** indoor environments

We also performed some experiments for the scenario in which multiple witness devices participate in the LP generation process. To evaluate the effect of device mobility, we performed the outdoor experiments while the prover and witness devices were moving with an average speed of 1.2 m/s. Figure 4.33 shows the 300 m outdoor path that we used for the mobility test. During the mobility test, an average distance of 7 m was maintained between the prover and witness devices. Figure 4.34a and b show the time required by five different witness devices to generate LPs for a single prover device in indoor and outdoor environments, respectively. We noticed an average increase of 8% in the latency of LP generation for the indoor environment. This is due to signal attenuations, absorptions and reflections caused by indoor elements such as walls, windows, and furniture. However, it does not have a significant impact on the system performance. Therefore, PASPORT performs well in indoor environments. It is expected that PASPORT shows a better performance if users communicate using WiFi as it provides more coverage distance than Bluetooth.

References

1. Y. Qu, S. Yu, L. Gao, W. Zhou, S. Peng, A hybrid privacy protection scheme in cyber-physical social networks. IEEE Trans. Comput. Social Syst. **5**(3), 773–784 (2018)
2. M.U. Hassan, M.H. Rehmani, J. Chen, Differential privacy techniques for cyber physical systems: a survey. IEEE Commun. Surv. Tutor. **22**(1), 746–789 (2019)
3. S. Yu, M. Liu, W. Dou, X. Liu, S. Zhou, Networking for big data: a survey. IEEE Commun. Surv. Tutor. **19**(1), 531–549 (2017)
4. W. Wang, Q. Zhang, Privacy preservation for context sensing on smartphone. IEEE/ACM Trans. Netw. **24**(6), 3235–3247 (2016)
5. B.S. Gu, L. Gao, X. Wang, Y. Qu, J. Jin, S. Yu, Privacy on the edge: customizable privacy-preserving context sharing in hierarchical edge computing. IEEE Trans. Netw. Sci. Eng. (2019)
6. Y. Qu, L. Gao, T.H. Luan, Y. Xiang, S. Yu, B. Li, G. Zheng, Decentralized privacy using blockchain-enabled federated learning in fog computing. IEEE Inter. Things J. (2020)
7. J. Xiong, R. Ma, L. Chen, Y. Tian, Q. Li, X. Liu, Z. Yao, A personalized privacy protection framework for mobile crowdsensing in iiot. IEEE Trans. Industr. Inf. **16**(6), 4231–4241 (2019)
8. M. Rezvani, A. Ignjatovic, E. Bertino, S. Jha, Secure data aggregation technique for wireless sensor networks in the presence of collusion attacks. IEEE Trans. Dependable Secure Comput. **12**(1), 98–110 (2015)
9. Y. Zhao, I. Wagner, Using metrics suites to improve the measurement of privacy in graphs. IEEE Trans. Dependable Secure Comput. (2020)
10. L. Qi, X. Zhang, S. Li, S. Wan, Y. Wen, W. Gong, Spatial-temporal data-driven service recommendation with privacy-preservation. Inf. Sci. **515**, 91–102 (2020)
11. K. Xu, Y. Guo, L. Guo, Y. Fang, X. Li, My privacy my decision: Control of photo sharing on online social networks. IEEE Trans. Dependable Secure Comput. **14**(2), 199–210 (2017)
12. Y. Qu, S. Yu, W. Zhou, Y. Tian, Gan-driven personalized spatial-temporal private data sharing in cyber-physical social systems. IEEE Trans. Netw. Sci. Eng. (2020)
13. A.R. Sfar, E. Natalizio, Y. Challal, Z. Chtourou, A roadmap for security challenges in the internet of things. Digit. Commun. Netw. **4**(2), 118–137 (2018)
14. S. Pierangela, S. Latanya, Protecting privacy when disclosing information: k-anonymity and its enforcement through generalization and suppression, in *Proceedings of the IEEE Symposium on Research in Security and Privacy* (1998), pp. 1–19
15. A. Machanavajjhala, D. Kifer, J. Gehrke, M. Venkitasubramaniam, *L*-diversity: privacy beyond *k*-anonymity. IEEE Trans. Knowl. Data Eng. **1**(1) (2007)
16. N. Li, T. Li, S. Venkatasubramanian, Closeness: a new privacy measure for data publishing. IEEE Trans. Knowl. Data Eng. **22**(7), 943–956 (2010)
17. C. Dwork, Differential privacy, in *Proceedings of ICALP 2006, Venice, Italy, July 10-14, 2006, Proceedings, Part II* (2006), pp. 1–12
18. C. Dwork, F. McSherry, K. Nissim, A.D. Smith, Calibrating noise to sensitivity in private data analysis, in *Theory of Cryptography, Third Theory of Cryptography Conference TCC, New York, NY, USA, March 4–7* (2006), pp. 265–284
19. T. Zhu, G. Li, W. Zhou, P.S. Yu, Differentially private data publishing and analysis: a survey. IEEE Trans. Knowl. Data Eng. **29**(8), 1619–1638 (2017)
20. S. Han, S. Zhao, Q. Li, C. Ju, W. Zhou, PPM-HDA: privacy-preserving and multifunctional health data aggregation with fault tolerance. IEEE Trans. Inf. Forensics Secur. **11**(9), 1940–1955 (2016)
21. R. Yu, J. Kang, X. Huang, S. Xie, Y. Zhang, S. Gjessing, Mixgroup: accumulative pseudonym exchanging for location privacy enhancement in vehicular social networks. IEEE Trans. Dependable Secure Comput. **13**(1), 93–105 (2016)
22. Y. Qu, S. Yu, W. Zhou, S. Peng, G. Wang, K. Xiao, Privacy of things: Emerging challenges and opportunities in wireless internet of things. IEEE Wirel. Commun. **25**(6), 91–97 (2018)
23. X. Zhang, M.M. Khalili, M. Liu, Recycled admm: improving the privacy and accuracy of distributed algorithms. IEEE Trans. Inf. Forensics Secur. **15**, 1723–1734 (2019)

24. C.X. Wang, Y. Song, W.P. Tay, Arbitrarily strong utility-privacy tradeoff in multi-agent systems. IEEE Trans. Inf. Forensics Secur. (2020)
25. M. Sun, K. Sakai, W. Ku, T. Lai, A.V. Vasilakos, Private and secure tag access for large-scale RFID systems. IEEE Trans. Dependable Secure Comput. **13**(6), 657–671 (2016)
26. Z. Jorgensen, T. Yu, G. Cormode, Conservative or liberal? Personalized differential privacy, in *2015 IEEE 31st International Conference on Data Engineering (ICDE)*. IEEE (2015), pp. 1023–1034
27. J. Yu, K. Wang, D. Zeng, C. Zhu, S. Guo, Privacy-preserving data aggregation computing in cyber-physical social systems. ACM Trans. Cyber-Phys. Syst. **3**(1), 1–23 (2018)
28. M. Keshk, E. Sitnikova, N. Moustafa, J. Hu, I. Khalil, An integrated framework for privacy-preserving based anomaly detection for cyber-physical systems. IEEE Trans. Sustain. Comput. (2019)
29. F. Koufogiannis, G.J. Pappas, Diffusing private data over networks. IEEE Trans. Control Netw. Syst. (2016)
30. P. Wijesekera, A. Baokar, L. Tsai, J. Reardon, S. Egelman, D.A. Wagner, K. Beznosov, The feasibility of dynamically granted permissions: aligning mobile privacy with user preferences, in *IEEE Symposium on Security and Privacy, SP 2017, San Jose, CA, USA, May 22–26* (2017), pp. 1077–1093
31. F. Amiri, N. Yazdani, A. Shakery, S.S. Ho, Bayesian-based anonymization framework against background knowledge attack in continuous data publishing. Trans. Data Priv. **12**(3), 197–225 (2019)
32. Y. Liu, N. Li, Retrieving hidden friends: a collusion privacy attack against online friend search engine. IEEE Trans. Inf. Forensics Secur. **14**(4), 833–847 (2018)
33. J. Jiang, W. Sheng, S. Yu, Y. Xiang, W. Zhou, Rumor source identification in social networks with time-varying topology. IEEE Trans. Dependable Secure Comput. (2016)
34. S. Ma, K. Feng, J. Li, H. Wang, G. Cong, J. Huai, Proxies for shortest path and distance queries. IEEE Trans. Knowl. Data Eng. **28**(7), 1835–1850 (2016)
35. C. Aporntewan, P. Chongstitvatana, N. Chaiyaratana, Indexing simple graphs by means of the resistance distance. IEEE Access **4**, 5570–5578 (2016)
36. S.P. Kasiviswanathan, K. Nissim, S. Raskhodnikova, A. Smith, Analyzing graphs with node differential privacy, in *Theory of Cryptography*. (Springer, Berlin, 2013), pp. 457–476
37. Q. Xiao, R. Chen, K.-L. Tan, Differentially private network data release via structural inference, in *Proceedings of the 20th ACM SIGKDD International Conference on Knowledge Discovery and Data Mining*. ACM (2014), pp. 911–920
38. G. Zhang, X. Liu, Y. Yang, Time-series pattern based effective noise generation for privacy protection on cloud. IEEE Trans. Comput. **64**(5), 1456–1469 (2015)
39. R. Liu, J. Cao, K. Zhang, W. Gao, J. Liang, L. Yang, When privacy meets usability: unobtrusive privacy permission recommendation system for mobile apps based on crowdsourcing. IEEE Trans. Serv. Comput. (2016)
40. C. Dwork, Differential privacy, in *Encyclopedia of Cryptography and Security*, 2nd edn. (2011), pp. 338–340
41. E.L. Lehmann, G. Casella, *Theory of Point Estimation*. (Springer Science & Business Media, Berlin, 2006)
42. H. Song, G. Fink, S. Jeschke, *Security and Privacy in Cyber-Physical Systems*. (Wiley Online Library, 2017)
43. J. McAuley, J. Leskovec, Social circles: Google+, https://snap.stanford.edu/data/egonets-Gplus.html
44. J. Guare, *Six Degrees of Separation: A Play* (Vintage, New York, 1990)
45. W. Shu, Y.-H. Chuang, The perceived benefits of six-degree-separation social networks. Internet Research (2011)
46. H. Kardes, A. Sevincer, M.H. Gunes, M. Yuksel, Six degrees of separation among us researchers, in *2012 IEEE/ACM International Conference on Advances in Social Networks Analysis and Mining*. IEEE (2012), pp. 654–659

47. Number of monthly active facebook users worldwide, https://www.statista.com/statistics/264810/number-of-monthly-active-facebook-users-worldwide/
48. E. Novak, Q. Li, Near-pri: private, proximity based location sharing, in *IEEE INFOCOM 2014 - IEEE Conference on Computer Communications* (2014), pp. 37–45
49. R. Schlegel, C. Chow, Q. Huang, D.S. Wong, Privacy-preserving location sharing services for social networks. IEEE Trans. Serv. Comput. **10**(5), 811–825 (2017)
50. M.R. Nosouhi, K. Sood, S. Yu, M. Grobler, J. Zhang, Pasport: a secure and private location proof generation and verification framework. IEEE Trans. Comput. Soc. Syst. **7**(2), 293–307 (2020)
51. M.R. Nosouhi, Y. Qu, S. Yu, Y. Xiang, D. Manuel, Distance-based location privacy protection in social networks, in *2017 27th International Telecommunication Networks and Applications Conference (ITNAC)* (2017), pp. 1–6
52. S. Ribeiro, G. Pappa, Strategies for combining twitter users geo-location methods. GeoInformatica **22**, 563–587 (2017)
53. O. Ajao, J. Hong, W. Liu, A survey of location inference techniques on twitter. **41**(6), 855–864 (2015). https://doi.org/10.1177/0165551515602847
54. A. Zubiaga, A. Voss, R. Procter, M. Liakata, B. Wang, A. Tsakalidis, Towards real-time, country-level location classification of worldwide tweets. IEEE Trans. Knowl. Data Eng. **29**(9), 2053–2066 (2017)
55. F. Alrayes, A. Abdelmoty, Towards location privacy awareness on geo-social networks," in *2016 10th International Conference on Next Generation Mobile Applications, Security and Technologies (NGMAST)* (2016), pp. 105–114
56. S. Wang, R. Sinnott, S. Nepal, Protecting the location privacy of mobile social media users, in *IEEE International Conference on Big Data (Big Data)* (2016), pp. 1143–1150
57. J.D. Zhang, G. Ghinita, C.Y. Chow, Differentially private location recommendations in geosocial networks, in *2014 IEEE 15th International Conference on Mobile Data Management*, vol. 1 (2014), pp. 59–68
58. X. Zheng, Z. Cai, J. Li, H. Gao, Location-privacy-aware review publication mechanism for local business service systems, in *IEEE INFOCOM 2017 - IEEE Conference on Computer Communications* (2017), pp. 1–9
59. J.H. Abawajy, M.I.H. Ninggal, T. Herawan, Privacy preserving social network data publication. IEEE Commun. Surv. Tutor. **18**(3), 1974–1997 (2016)
60. M. Fire, R. Goldschmidt, Y. Elovici, Online social networks: threats and solutions. IEEE Commun. Surv. Tutor. **16**(4), 2019–2036 (2014)
61. F. Koufogiannis, G.J. Pappas, Diffusing private data over networks. IEEE Trans. Control Netw. **PP**(99), 1–1 (2017)
62. M.E. Andrés, N.E. Bordenabe, K. Chatzikokolakis, C. Palamidessi, Geo-indistinguishability: differential privacy for location-based systems, in *Proceedings of the 2013 ACM SIGSAC Conference on Computer & Communications Security*. Association for Computing Machinery (2013), pp. 901–914. https://doi.org/10.1145/2508859.2516735
63. F. Koufogiannis, S. Han, G.J. Pappas, Gradual release of sensitive data under differential privacy. J. Priv. Confidentiality **7** (2016)
64. M.R. Nosouhi, V.V.H. Pham, S. Yu, Y. Xiang, M. Warren, A hybrid location privacy protection scheme in big data environment, in *GLOBECOM 2017 - 2017 IEEE Global Communications Conference* (2017), pp. 1–6
65. C. Ruiz Vicente, D. Freni, C. Bettini, C.S. Jensen, Location-related privacy in geo-social networks. IEEE Int. Comput. **15**(3), 20–27 (2011)
66. E. ElSalamouny, S. Gambs, Differential privacy models for location-based services. Trans. Data Privacy **9**(1), 15–48 (2016). (Apr.)
67. A.I. Abdelmoty, F. Alrayes, Towards understanding location privacy awareness on geo-social networks. ISPRS Int. J. Geo-Inf **109**(1), 15–48 (2017)
68. M. Andrés, N. Bordenabe, K. Chatzikokolakis, C. Palamidessi, Geo-indistinguishability: differential privacy for location-based systems. CCS Assoc. Comput. Mach. **2013**, 901–914 (2013)

69. C. Dwork, A. Roth, The algorithmic foundations of differential privacy. Found. Trends Theor. Comput. Sci. **9**(3–4), 211–407 (2014)
70. J. Jiang, S. Wen, S. Yu, Y. Xiang, W. Zhou, K-center: An approach on the multi-source identification of information diffusion. IEEE Trans. Inf. Forensics Secur. **10**(12), 2616–2626 (2015)
71. P. Welke, A. Markowetz, T. Suel, M. Christoforaki, Three-hop distance estimation in social graphs, in *IEEE International Conference on Big Data (Big Data)* (2016), pp. 1048–1055
72. J. Cheng, Y. Zhang, Q. Ye, H. Du, High-precision shortest distance estimation for large-scale social networks, in *IEEE INFOCOM 2016 - The 35th Annual IEEE International Conference on Computer Communications* (2016), pp. 1–9
73. E.W. Stacy, A generalization of the gamma distribution. Ann. Math. Stat. **33**(3), 1187–1192 (1962)
74. D. Goldschlag, M. Reed, P. Syverson, Onion routing for anonymous and private internet connections (1999)
75. O. Berthold, H. Federrath, S. Köpsell, Web mixes: a system for anonymous and unobservable internet access, in *Designing Privacy Enhancing Technologies* (2000), pp. 115–129
76. R. Dingledine, N. Mathewson, P. Syverson, Tor: The second-generation onion router, in *Proceedings of the 13th Conference on USENIX Security Symposium - Volume 13*. USENIX Association (2004), p. 21
77. D.L. Chaum, Untraceable electronic mail, return addresses, and digital pseudonyms. Commun. ACM **24**(2), 84–90 (1981). https://doi.org/10.1145/358549.358563
78. M.R. Nosouhi, S. Yu, K. Sood, M. Grobler, Hsdc–net: secure anonymous messaging in online social networks, in *2019 18th IEEE International Conference On Trust, Security And Privacy In Computing And Communications/13th IEEE International Conference On Big Data Science And Engineering (TrustCom/BigDataSE)* (2019), pp. 350–357
79. S. Angel, S. Setty, Unobservable communication over fully untrusted infrastructure, in *12th USENIX Symposium on Operating Systems Design and Implementation (OSDI 16)*. USENIX Association (2016), pp. 551–569, https://www.usenix.org/conference/osdi16/technical-sessions/presentation/angel
80. H. Corrigan-Gibbs, D.I. Wolinsky, B. Ford, Proactively accountable anonymous messaging in verdict, in *22nd USENIX Security Symposium (USENIX Security 13)*. USENIX Association (2013), pp. 147–162, https://www.usenix.org/conference/usenixsecurity13/technical-sessions/presentation/corrigan-gibbs
81. A. Kwon, D. Lazar, S. Devadas, B. Ford, Riffle: an efficient communication system with strong anonymity. Proc. Priv. Enhancing Technol. **2016**(2), 115–134 (2016). https://doi.org/10.1515/popets-2016-0008
82. S.J. Murdoch, G. Danezis, Low-cost traffic analysis of tor, in *2005 IEEE Symposium on Security and Privacy (S P'05)* (2005), pp. 183–195
83. D. Chaum, The dining cryptographers problem: unconditional sender and recipient untraceability. J. Cryptol. **1**, 65–75 (1988)
84. H. Corrigan-Gibbs, B. Ford, Dissent: Accountable anonymous group messaging, in *Proceedings of the 17th ACM Conference on Computer and Communications Security*, ser. CCS '10. Association for Computing Machinery (2010), pp. 340–350. https://doi.org/10.1145/1866307.1866346
85. S. Goel, M. Robson, M. Polte, E. Sirer, Herbivore: a scalable and efficient protocol for anonymous communication. Technical Report, Cornell University (2003)
86. D.I. Wolinsky, H. Corrigan-Gibbs, B. Ford, A. Johnson, Dissent in numbers: making strong anonymity scale, in *10th USENIX Symposium on Operating Systems Design and Implementation (OSDI 12)*. USENIX Association (2012), pp. 179–182, https://www.usenix.org/conference/osdi12/technical-sessions/presentation/wolinsky
87. C.-Y. Chow, M.F. Mokbel, X. Liu, A peer-to-peer spatial cloaking algorithm for anonymous location-based service, in *Proceedings of the 14th Annual ACM International Symposium on Advances in Geographic Information Systems*, ser. GIS '06. Association for Computing Machinery (2006), pp. 171–178. https://doi.org/10.1145/1183471.1183500

88. P. Kalnis, G. Ghinita, K. Mouratidis, D. Papadias, Preventing location-based identity inference in anonymous spatial queries. IEEE Trans. Knowl. Data Eng. **19**(12), 1719–1733 (2007)
89. T. Xu, Y. Cai, Exploring historical location data for anonymity preservation in location-based services, in *IEEE INFOCOM 2008 - The 27th Conference on Computer Communications* (2008), pp. 547–555
90. Y. Wang, D. Xu, X. He, C. Zhang, F. Li, B. Xu, L2p2: location-aware location privacy protection for location-based services, in *2012 Proceedings IEEE INFOCOM* (2012), pp. 1996–2004
91. H. Kido, Y. Yanagisawa, T. Satoh, An anonymous communication technique using dummies for location-based services, in *ICPS '05. Proceedings. International Conference on Pervasive Services*(2005), pp. 88–97
92. H. Lu, C.S. Jensen, M.L. Yiu, Pad: privacy-area aware, dummy-based location privacy in mobile services, in *Proceedings of the Seventh ACM International Workshop on Data Engineering for Wireless and Mobile Access*, ser. MobiDE '08. Association for Computing Machinery (2008), pp. 16–23. https://doi.org/10.1145/1626536.1626540
93. B. Niu, Q. Li, X. Zhu, G. Cao, H. Li, Achieving k-anonymity in privacy-aware location-based services, in *IEEE INFOCOM 2014 - IEEE Conference on Computer Communications* (2014), pp. 754–762
94. H. Liu, X. Li, H. Li, J. Ma, X. Ma, Spatiotemporal correlation-aware dummy-based privacy protection scheme for location-based services, in *IEEE INFOCOM 2017 - IEEE Conference on Computer Communications* (2017), pp. 1–9
95. P. Golle, A. Juels, Dining cryptographers revisited, in *Advances in Cryptology - EUROCRYPT 2004*. ed. by C. Cachin, J.L. Camenisch (Springer, Berlin, 2004), pp. 456–473
96. www.isi.deterlab.net/
97. X. Zeng, S.K. Garg, P. Strazdins, P.P. Jayaraman, D. Georgakopoulos, R. Ranjan, Iotsim: A simulator for analysing iot applications. J. Syst. Arch. - Embedded Syst. Design **72**, 93–107 (2017)
98. X. Cheng, F. Lyu, W. Quan, C. Zhou, H. He, W. Shi, X. Shen, Space/aerial-assisted computing offloading for iot applications: A learning-based approach. IEEE J. Sel. Areas Commun. **37**(5), 1117–1129 (2019)
99. M.R. Alam, M.B.I. Reaz, M.A.M. Ali, A review of smart homes 2014;past, present, and future. IEEE Trans. Syst. Man Cybern. Part C (Applications and Reviews) **42**(6), 1190–1203 (2012)
100. A review of internet of things for smart home: Challenges and solutions. J. Clea. Prod. **140**(Part 3), 1454–1464 (2017)
101. A review of smart homes–present state and future challenges. Comput. Meth. Prog. Biomed. **91**(1), 55–81 (2008)
102. S.K. Datta, C. Bonnet, A. Gyrard, R.P.F. da Costa, K. Boudaoud, Applying internet of things for personalized healthcare in smart homes, in *2015 24th Wireless and Optical Communication Conference (WOCC)* (2015), pp. 164–169
103. F. Cicirelli, G. Fortino, A. Giordano, A. Guerrieri, G. Spezzano, A. Vinci, On the design of smart homes: a framework for activity recognition in home environment. J. Med. Syst. **40**(9), 200 (2016). (Jul)
104. Y. Jie, J.Y. Pei, L. Jun, G. Yun, X. Wei, Smart home system based on iot technologies, in *2013 International Conference on Computational and Information Sciences* (2013), pp. 1789–1791
105. A.V. Dastjerdi, R. Buyya, Fog computing: helping the internet of things realize its potential. Computer **49**(8), 112–116 (2016). (Aug)
106. T.H. Luan, L. Gao, Z. Li, Y. Xiang, L. Sun, Fog computing: focusing on mobile users at the edge (2015). arXiv:1502.01815
107. M. Chiang, T. Zhang, Fog and iot: an overview of research opportunities. IEEE Int. Things J. **3**(6), 854–864 (2016). (Dec)
108. A. Brogi, S. Forti, Qos-aware deployment of iot applications through the fog. IEEE Int. Things J. **4**(5), 1185–1192 (2017). (Oct)

109. B. Tang, Z. Chen, G. Hefferman, T. Wei, H. He, Q. Yang, A hierarchical distributed fog computing architecture for big data analysis in smart cities, in *Proceedings of the ASE BigData & SocialInformatics 2015*, ser. ASE BD&SI '15 (2015), pp. 28:1–28:6

110. S.K. Datta, C. Bonnet, J. Haerri, Fog computing architecture to enable consumer centric internet of things services, in *2015 International Symposium on Consumer Electronics (ISCE)* (2015), pp. 1–2

111. W. Lee, K. Nam, H.G. Roh, S.H. Kim, A gateway based fog computing architecture for wireless sensors and actuator networks, in *2016 18th International Conference on Advanced Communication Technology (ICACT)* (2016), pp. 210–213

112. Y. Qu, S. Yu, L. Gao, J. Niu, Big data set privacy preserving through sensitive attribute-based grouping, in *IEEE International Conference on Communications, ICC 2017, Paris, France, May 21–25* (2017), pp. 1–6

113. C. Dwork, K. Kenthapadi, F. McSherry, I. Mironov, M. Naor, Our data, ourselves: privacy via distributed noise generation, in *Advances in Cryptology - EUROCRYPT 2006, 25th Annual International Conference on the Theory and Applications of Cryptographic Techniques, St. Petersburg, Russia, May 28 - June 1, 2006, Proceedings* (2006), pp. 486–503

114. T. Song, R. Li, B. Mei, J. Yu, X. Xing, X. Cheng, A privacy preserving communication protocol for iot applications in smart homes. IEEE Int. Things J. **4**(6), 1844–1852 (2017)

115. Y. Lee, W. Hsiao, Y. Lin, S.T. Chou, Privacy-preserving data analytics in cloud-based smart home with community hierarchy. IEEE Trans. Consum. Electron. **63**(2), 200–207 (2017)

116. H. Li, H. Zhu, S. Du, X. Liang, X. Shen, Privacy leakage of location sharing in mobile social networks: Attacks and defense. IEEE Trans. Dependable Secure Comput. (2016)

117. S. Yu, Big privacy: challenges and opportunities of privacy study in the age of big data. IEEE Access **4**, 2751–2763 (2016)

118. Y. Qu, S. Yu, L. Gao, W. Zhou, S. Peng, A hybrid privacy protection scheme in cyber-physical social networks. IEEE Trans. Comput. Soc. Syst. **5**(3), 773–784 (2018)

119. N. Komninos, E. Philippou, A. Pitsillides, Survey in smart grid and smart home security: issues, challenges and countermeasures. IEEE Commun. Surv. Tutori. **16**(4), 1933–1954, Fourthquarter (2014)

120. D. Geneiatakis, I. Kounelis, R. Neisse, I. Nai-Fovino, G. Steri, G. Baldini, Security and privacy issues for an iot based smart home, in *2017 40th International Convention on Information and Communication Technology, Electronics and Microelectronics (MIPRO)* (2017), pp. 1292–1297

121. K. Lee, D. Kim, D. Ha, U. Rajput, H. Oh, On security and privacy issues of fog computing supported internet of things environment, in *2015 6th International Conference on the Network of the Future (NOF)* (2015), pp. 1–3

122. M. Yang, T. Zhu, T. Ma, Y. Xiang, W. Zhou, Privacy preserving collaborative filtering via the johnson-lindenstrauss transform, in *2017 IEEE Trustcom/BigDataSE/ICESS, Sydney, Australia, August 1–4* (2017), pp. 417–424

123. K. Zhang, J. Ni, K. Yang, X. Liang, J. Ren, X.S. Shen, Security and privacy in smart city applications: challenges and solutions. IEEE Commun. Mag. **55**(1), 122–129 (2017). (January)

124. X. Gong, X. Chen, K. Xing, D. Shin, M. Zhang, J. Zhang, Personalized location privacy in mobile networks: a social group utility approach, in *2015 IEEE Conference on Computer Communications, INFOCOM 2015, Kowloon, Hong Kong, April 26 - May 1* (2015), pp. 1008–1016

125. Y. Qu, S. Yu, W. Zhou, S. Peng, G. Wang, K. Xiao, Privacy of things: emerging challenges and opportunities in wireless internet of things. IEEE Wireless Commun. **25**(6), 91–97 (2018)

126. M. Götz, S. Nath, J. Gehrke, Maskit: privately releasing user context streams for personalized mobile applications, in *Proceedings of the ACM SIGMOD International Conference on Management of Data, SIGMOD 2012, Scottsdale, AZ, USA, May 20–24* (2012), pp. 289–300

127. E. Aghasian, S.K. Garg, L. Gao, S. Yu, J. Montgomery, Scoring users' privacy disclosure across multiple online social networks. IEEE Access **5**, 13 118–13 130 (2017)

128. A. Fleury, M. Vacher, N. Noury, Svm-based multimodal classification of activities of daily living in health smart homes: sensors, algorithms, and first experimental results. IEEE Trans. Inf Technol. Biomed. **14**(2), 274–283 (2010)

129. A. Fleury, N. Noury, M. Vacher, Improving supervised classification of activities of daily living using prior knowledge, in *Digital Advances in Medicine, E-Health, and Communication Technologies*(2013), p. 131

130. M.R. Nosouhi, S. Yu, W. Zhou, M. Grobler, H. Keshtiar, Blockchain for secure location verification. J. Parall. Distr. Comput. **136**, 40–51 (2020)

131. M.R. Nosouhi, S. Yu, M. Grobler, Q. Zhu, Y. Xiang, Blockchain–based location proof generation and verification, in *IEEE INFOCOM 2019 - IEEE Conference on Computer Communications Workshops (INFOCOM WKSHPS)* (2019), pp. 1–6

132. Z. Zhang, L. Zhou, X. Zhao, G. Wang, Y. Su, M. Metzger, H. Zheng, B.Y. Zhao, On the validity of geosocial mobility traces, in *Proceedings of the Twelfth ACM Workshop on Hot Topics in Networks*, ser. HotNets-XII. Association for Computing Machinery (2013). https://doi.org/10.1145/2535771.2535786

133. Z. Gao, H. Zhu, Y. Liu, M. Li, Z. Cao, Location privacy in database-driven cognitive radio networks: attacks and countermeasures, in *Proceedings IEEE INFOCOM* (2013), pp. 2751–2759

134. B. Waters, E. Felten, Secure, private proofs of location. Technical Report, Department of Computer Science, Princeton University (2003)

135. S. Saroiu, A. Wolman, Enabling new mobile applications with location proofs, in *Proceedings of the 10th Workshop on Mobile Computing Systems and Applications*, ser. HotMobile '09. Association for Computing Machinery (2009). https://doi.org/10.1145/1514411.1514414

136. C. Javali, G. Revadigar, K.B. Rasmussen, W. Hu, S. Jha, I am alice, i was in wonderland: Secure location proof generation and verification protocol, in *2016 IEEE 41st Conference on Local Computer Networks (LCN)* (2016), pp. 477–485

137. W. Luo, U. Hengartner, Veriplace: a privacy-aware location proof architecture, in *GIS '10* (2010)

138. X. Wang, A. Pande, J. Zhu, P. Mohapatra, Stamp: Enabling privacy-preserving location proofs for mobile users. IEEE/ACM Trans. Netw. **24**(6), 3276–3289 (2016)

139. Z. Zhu, G. Cao, Toward privacy preserving and collusion resistance in a location proof updating system. IEEE Trans. Mob. Comput. **12**(1), 51–64 (2013)

140. B. Davis, H. Chen, M. Franklin, Privacy-preserving alibi systems, in *Proceedings of the 7th ACM Symposium on Information, Computer and Communications Security*, ser. ASIACCS '12. Association for Computing Machinery (2012), pp. 34–35. https://doi.org/10.1145/2414456.2414475

141. M. Talasila, R. Curtmola, C. Borcea, Link: location verification through immediate neighbors knowledge, in *International Conference on Mobile and Ubiquitous Systems: Computing, Networking, and Services.* (Springer, 2010), pp. 210–223

142. M. Reza Nosouhi, S. Yu, M. Grobler, Y. Xiang, Z. Zhu, Sparse: Privacy-aware and collusion resistant location proof generation and verification, in *2018 IEEE Global Communications Conference (GLOBECOM)* (2018), pp. 1–6

143. S. Gambs, M. Killijian, M. Roy, M. Traoré, Props: a privacy-preserving location proof system, in *2014 IEEE 33rd International Symposium on Reliable Distributed Systems* (2014), pp. 1–10

144. L. Bussard, W. Bagga, Distance-bounding proof of knowledge to avoid real-time attacks, in *IFIP International Information Security Conference.* (Springer, 2005), pp. 223–238

145. I. Boureanu, S. Vaudenay, Challenges in distance bounding. IEEE Secur. Priv. **13**(1), 41–48 (2015)

146. M. Fischlin, C. Onete, Terrorism in distance bounding: modeling terrorist-fraud resistance, in *International Conference on Applied Cryptography and Network Security.* (Springer, 2013), pp. 414–431

147. C.H. Kim, G. Avoine, F. Koeune, F.-X. Standaert, O. Pereira, The swiss-knife rfid distance bounding protocol, in *International Conference on Information Security and Cryptology.* (Springer, 2008), pp. 98–115

148. S. Gambs, C. Onete, J.-M. Robert, Prover anonymous and deniable distance-bounding authentication, in *Proceedings of the 9th ACM Symposium on Information, Computer and Communications Security* (2014), pp. 501–506

149. C. Cremers, K.B. Rasmussen, B. Schmidt, S. Capkun, Distance hijacking attacks on distance bounding protocols, in *IEEE Symposium on Security and Privacy*. IEEE (2012), pp. 113–127

150. S. Vaudenay, Private and secure public-key distance bounding, in *International Conference on Financial Cryptography and Data Security*. (Springer, 2015), pp. 207–216

151. S. Vaudenay, I. Boureanu, A. Mitrokotsa, et al., Practical & provably secure distance-bounding, in *The 16th Information Security Conference*, vol. 10 (2013), pp. 978–3

152. G. Avoine, X. Bultel, S. Gambs, D. Gerault, P. Lafourcade, C. Onete, J.-M. Robert, A terrorist-fraud resistant and extractor-free anonymous distance-bounding protocol, in *Proceedings of the 2017 ACM on Asia Conference on Computer and Communications Security* (2017), pp. 800–814

153. A. Pham, K. Huguenin, I. Bilogrevic, I. Dacosta, J.-P. Hubaux, Securerun: cheat-proof and private summaries for location-based activities. IEEE Trans. Mob. Comput. **15**(8), 2109–2123 (2015)

Chapter 5
Future Research Directions

In previous chapters, we have present the big picture of existing personalized privacy protection solutions along with several privacy concerns and leading attacks. However, a mass of significant and prospective issues remain under-explored. The prosperity of machine learning, the Internet of Things, and blockchain brings appealing opportunities for researches on personalized privacy protection while posing further challenges in data utility upgradation in big data scenarios. Beyond this, there are numerous other topics that desiderata consideration in personalized privacy protection, and we outline several potentially promising research directions that may be worthy of future efforts.

5.1 Personalized Privacy-Preserving Attribute-based Encryption

Cryptography has always been a popular and promising tool for privacy protection. However, how to leverage cryptography-based methods to achieve personalized privacy protection is still in its infancy. Based on the authors' review and analysis of existing literature, we identify the feasibility of using attribute-based encryption (ABE) to provide privacy protection in a personalized manner.

Currently, the gap between ABE and personalized privacy protection is how to quantify the attributes of ABE, and further, how to may the quantified attributes to a specific privacy protection level. To close the gap, it is necessary to create an attribute-by-data matrix, in which all pre-defined attributes are the entities of each column while each row represents a piece of encrypted data. In the matrix, the value is either 1 or 0. 1 denotes the corresponding attribute is necessary to decrypt this piece of data while 0 means this attribute is undesired. In this way, each row is transformed into a multi-dimensional attribute vector that can be measured and compared. Besides, to avoid the situation that several pieces of data require the same subset of attributes, we put a binary-number index to denote the index vector of each piece of data before the attribute vector and thereby obtain a unique vector for each piece of data. Similarly, we can model an individual with several attributes as an attribute vector and compute the cross product with each of the rows in the attribute-by-data matrix. Then, we will keep the unchanged rows in the matrix as the personalized matrix for this individual. The construction of an attribute-by-data matrix can achieve quantifying and mapping simultaneously.

Using ABE to attain personalized privacy protection is a special case of access control. This type of personalized access control is different from the current binary access control in cryptography-based methods. It is more flexible and thereby can meet the diverse demands of real-world privacy protection.

5.2   Personalized Privacy-Preserving Federated Learning Using Generative Adversarial Network

In addition to cryptography-based methods, machine learning is playing an important role in privacy protection. On one hand, some machine learning algorithms may reveal sensitive information due to its predictive features. On the other hand, several machine learning algorithms and novel paradigms are proposed to achieve privacy protection.

One of the most popular machine learning paradigms that aim at addressing privacy issues is federated learning. It enables the protection of local data and efficient communication simultaneously. However, the model parameters received by the central server of federated learning can be potentially accessed by others, which may be used to breach the local data privacy if a series of model parameters are collected. Although homomorphic encryption is a technique that allows a central server to operate on the ciphertext of model parameters, it highly relies on a very complicated and powerful computing infrastructure, which is not feasible in the proposed edge intelligence platform.

To preserve the privacy of model parameters, differential privacy (DP) and its extensions are the mainstream methods. DP aims to provide strict privacy protection to hide sensitive information by adding controllable noise. However, the existing DP-based solutions hold the common assumption that the injected noise complies with the Laplace mechanism. It usually does not hold for federated learning, in which the generated data is too sparse to follow a specific distribution. Moreover, the assumption of Laplace mechanism compliance has a significantly and randomly negative impact on data utility. As a result, the direct deployment of DP to the model parameters will drag down the performance of federated learning or even cause a failure of convergence. It is shown that GAN could be used to generate imitative synthetic data to potentially protect the privacy of sensitive data constrained by game theory. However, it cannot provide a strict privacy guarantee with solid theory, which fails to meet the privacy requirements of edge computing.

Therefore, we plan to design a generative adversarial network (GAN) enhanced DP scheme to achieve personalized privacy-preserving federated learning while guaranteeing high data utility. In order to develop the personalized GAN enhanced DP algorithm, we will first investigate the feasibility of using GAN to generate differentially private synthetic model parameters, for which a new perceptron is proposed, namely, DP-Identifier. Then the DP-Identifier is designed to be part of the game process to model the confrontation with the Generator and Discriminator, which are two key components of a classic GAN. In this case, the Generator generates synthetic model parameters while the Discriminator and DP-Identifier try to check if the generated model parameters are not distinguishable from origin model parameters and satisfy DP requirements simultaneously. The two correlated gaming processes are constrained and optimized by game theory with iteration. Subsequently, the generated synthetic model parameters comply with DP requirements without the impact of sparse data while maximizing the data utility by removing the high randomness caused by the Laplace mechanism. Furthermore, we plan to extend the current static game process in GAN to a dynamic one that considers the impact on the convergence and system status caused by adversaries' actions. The dynamic game process can be represented by an extension of the Markov decision process. In this way, the novel DP mechanism can adaptively adjust the balance between privacy protection and model parameters utility which are controlled by the DP-Identifier and the Discriminator, respectively. The novel DP mechanism is able to protect the privacy of the model parameters in federated learning while not dragging down the performances and convergence speed. New cost functions and adversary models should be proposed and optimized correspondingly.

5.3   Personalized Privacy-Preserving Blockchain-Enabled Federated Learning

As mentioned in the previous section, personalized privacy-preserving federated learning can be enabled using GAN. However, it still uses the traditional centralized processing FL paradigm, which leads to single-point failure and man-in-the-middle attacks. Besides, it faces several other challenges, such as data falsification and lack of incentive mechanism. Therefore, we intend to devise a novel FL paradigm, in particular, blockchain-enabled federated learning (BE-FL).

Firstly, it is essential to describe the devised consensus algorithm, namely, Proof-of-Federated-Learning (PoFL), and how the federated learning process is interpreted by PoFL. PoFL is the backbone of the blockchain-enabled federated learning architecture. In addition, we propose a novel personalized incentive mechanism, which motivates the participation of consensus and federated learning tasks simultaneously. In this paradigm, we use participants to denote the entities taking part in the training process in classic FL and miner to denote all the participants of the blockchain-enabled federated learning system. To achieve higher performance, novel mechanisms are deployed to select a group of local models. The miners associated with the selected local models are defined as the training miners who are involved in both the FL training and consensus processes. Besides, we name the miner who generates the candidate block as the aggregation miner. In the global aggregation phase, all training miners can conduct the aggregation and broadcast their blocks containing the corresponding global parameters. Each of the training miners will select the global parameters which are the closest to his/her own during a time threshold. After reaching the time threshold, the block with the most-selected global parameters is the candidate block while the associated miner is the aggregation miner of this round.

In PoFL, the mining process of classical blockchain systems corresponds to the local model training while the block generation and propagation process correspond to local model selection and global model aggregation in FL. The whole workflow starts with the first-round local model training, after which trained parameters of all local models are saved in blocks and then broadcast to the blockchain network. A smart contract, which is forced to execute by all BE-FL miners, deploys an improved detection algorithm of falsified local model parameters and an optimal local model selection algorithm. The detection algorithm filters out all the falsified local model parameters and the optimal local model selection algorithm identifies the local models that are best for global aggregation. Then the training miners have the right to train the selected local model parameters and generate a set of global model parameters with an advanced aggregation method. The shared cost functions are used to determine the quality of the local models, with which the local model parameters are selected with dynamic optimization algorithms. The training miners, as mentioned above, then generates a set of the global model parameters and storing them in a candidate block, which is again broadcast to the blockchain network. After cross verification, the candidate block with authenticated global model parameters will be appended on the ledger of each miner. After a defined threshold of time or percentage of miners is reached, the consensus process completes while the generator of the candidate block is regarded as the aggregation miner. The aggregation miner and training miners associated with all selected training local models are rewarded with a personalized incentive mechanism. Meanwhile, the miners use the global model parameters and their local data as the inputs to generate augmented synthetic data for the local training of the new round, which can mitigate heterogeneous issues. The iteration keeps going until the convergence of FL is reached. The personalized privacy protection is partially enabled by blockchain and partially achieved by FL, respectively.

5.4   Collusion Attack Resistance in Personalized Privacy Protection

Although personalized privacy protection provides many advantageous features, there are some emerging issues corresponding to it. One of the most severe issues is the collusion attack. As is known to all, a collusion attack is a traditional attack in the privacy-preserving domain. In this attack, multiple adversaries collude with each other with background knowledge owned by themselves. In personalized privacy protection scenarios, it becomes even risky since the published data received by each individual may be different. The difference leads to an extra possibility of privacy leakage.

For example, in differential privacy, personalized privacy means assigning different ϵ values to different individuals. Therefore, the published data satisfies $\{\epsilon_1, \epsilon_2, ..., \epsilon_n\}$-differential privacy. The composition mechanism, as a built-in mechanism in differential privacy, defines that if a ϵ_1-differential privacy mechanism and a ϵ_2-differential privacy mechanism composed with each other, the ceiling of the new mechanism satisfies by $(\epsilon_1 + \epsilon_2)$-differential privacy. Since a higher value of ϵ means a lower privacy protection level, the composition mechanism of differential privacy results in significant collusion attacks in personalized differential privacy.

One way to solve this issue is to update the noise generation mechanism. To deal with real-valued scenarios, the Laplace mechanism is widely used and proves its feasibility in real-world cases. It is possible to modify it such that the new Laplace mechanism generates the noise complying with a Markov stochastic process. In this way, the correlations of noise are de-coupled and therefore the composition mechanism works in a different way. The composed mechanism of ϵ_1-differential privacy and ϵ_2-differential privacy satisfies $\max(\epsilon_1, \epsilon_2)$-differential privacy. This will eliminate the incentive of the individual to collude with each other. Subsequently, the collusion attack can be defeated. A similar methodology may apply to other personalized privacy protection scenarios as well.

5.5   Trade-Off Optimization between Personalized Privacy Protection and Data Utility

Personalized privacy protection provides flexible privacy protection and enables a generalized privacy protection paradigm. However, in addition to personalized privacy protection, it is also essential to take data utility into consideration. Thus, another promising research direction is to optimize the trade-off between personalized privacy protection and data utility.

Data utility and personalized privacy protection can be both regarded as two key indexes of quality of service (QoS). Therefore, to achieve optimization, it is intuitive to create a QoS-based function that considers both of them. To model the problem, it is possible to use the Markov decision process (MDP). We can model the privacy protection levels as information loss, which is defined by information theory. Then the data utility can be measured by root-mean-square-error. Actually, all other measurements work, but we just discuss the most popular one here. Actions in MDP can be modeled by the values of privacy protection level, for example, the values of ϵ of differential privacy or the granularity of published data. States in MDP can be denoted by the published data of the current time slot and the attack response from the previous time slot. The rewards are then formulated as the data utility. By establishing a state transmission matrix, the MDP can dynamically optimize the trade-off considering the overall rewards of future time slots. Besides, there are other optimization methods, for instance, convex optimization, genetic algorithms, etc.

Chapter 6
Summary and Outlook

In this book, we summarize the latest work on personalized privacy protection in terms of information technology. Personalized privacy protection is still in its infancy. The theories, algorithms, and other conceptual designs surveyed in this book could be a starting point for forthcoming researchers and readers to probe this under-explored domain. We aim to offer a systematic summary of existing research and application outputs on personalized privacy protection, which also testifies the theoretical and practical applicability in diverse big data scenarios. We also subsequently present a couple of potentially promising directions, with which we expect to assist in avoiding superfluous efforts from subsequent interested explorers.

In general, an abundant volume of literature has been reviewed and analysed to show the current research and application status of personalized privacy protection solutions. We describe and compare the primary privacy concerns and attacks, some of which remain a bottleneck to personalized privacy protection. In particular, we have discussed the personalized privacy protection in cyber physical systems, social networks, smart homes, and location-based services. However, the proposed models are generalized models and able to be applied in more extensive scenarios.

We also include several mainstream theories for personalized privacy protection, including differential privacy, machine learning, game theory, and anonymity and clustering-based methods, and correspondingly explained and articulated while their feasibility has been demonstrated when fitting into various real-world practices.

Based on the existing research and results, we further discuss several future research directions, which are personalized privacy-preserving attribute-based encryption, personalized privacy-preserving federated learning using generative adversarial network, personalized privacy-preserving blockchain-enabled federated learning, collusion attack resistance in personalized privacy protection, and trade-off optimization between personalized privacy protection and data utility.

As aforementioned, privacy protection in the digital space is a new research domain. We have far more questions than answers, we definitely will counter many

© The Author(s), under exclusive license to Springer Nature Singapore Pte Ltd. 2021 137
Y. Qu et al., *Personalized Privacy Protection in Big Data*, Data Analytics,
https://doi.org/10.1007/978-981-16-3750-6_6

unprecedented problems, and many unknown of unknowns. Based on our study, we would like to share some big pictures with energetic young researcher as below.

First of all, we believe privacy protection needs the effort from multiple disciplines, for example, psychology, social science, law, information technology, and so on. It is certain that computer science herself cannot solve the privacy problem alone. According to the logic of science, we need firstly measure privacy, then represent privacy using mathematical models, and then confirm the ideas and conclusions by theoretical or experimental proof. Up to date, we do not have an effective way to get the first step done, namely measuring privacy. Similar to other soft concepts like happiness, madness, privacy is hard to measure. Issac Newton complained that "I can calculate the movement of stars, but not the madness of men", after 300 years, we still not face the similar difficulty. At the other hand, we see the dramatic development of all disciplines in the science family, we believe the light is on the horizon for us, but we do need to master multiple necessary skill sets to complete the mission.

Cross discipline research looks beautiful, but not easy to carry out. The Science magazine had a statistics several years ago, the result showed that among the research papers published on Science in the last 100 years, nearly 70% papers are cross disciplinary work. This result demonstrates that cross disciplinary study is powerful in research. However, our experience and the literature also show that it is hard to execute it. In general, a few coffees at the campus may generate some idea among colleagues from different discipline, but when we execute it, it is extremely hard as we speak different "languages". Strong leadership and financial support maybe the key for these kinds of collaboration. A common suggestion is we need to learn the skills of the other disciplines rather than bringing problems to the other party and waiting for solutions.

Secondly, theoretical tools for privacy is desperately needed, and it is a promising target for related communities. So far, differential privacy is the only new tool invented for privacy (we treat cryptography as the tool for secrecy sharing, not for privacy). However, differential privacy was invented for privacy protection in statistical information retrieval, which is only a very small part of the landscape of digital privacy. There are two possibilities on this issue.

- We extend the existing tools to deal with the new problems of privacy, e.g., upgrade cryptographic tools to fulfill the tasks of privacy preserving in big data publishing. The traditional symmetric and asymmetric encryption tools were designed to share a secret between two pairs (one-to-one communication), and the attribute-based encryption was developed to share a secret among a small group (one-to-many communication, we note the many here is a small number). However, in big data publishing, we release the data for anyone who wants to access under the condition of protecting the privacy of the data owners (one-to-any communication). So far cryptography cannot offer a suitable solution for it.
- Invent new tools. The invention of new tools is the result of application demands. We believe the research community will develop new tools for digital privacy as the demands are in place. It is a tough job, and also an exciting goal for hard working and talented people.

This small book is the short summary of our work in the recent years, and also the first step of our research group. We hope our shallow effort can attract interested readers to explore the promising land with us in both academia and industry domains.

We hope you enjoy reading the book, and sincerely looking forward to have your feedback, comments, suggestions and your work in the field.

Printed in the United States
by Baker & Taylor Publisher Services